Lecture Notes in Applied and Computational Mechanics

Volume 34

Series Editors

Prof. Dr.-Ing. Friedrich Pfeiffer
Prof. Dr.-Ing. Peter Wriggers

Lecture Notes in Applied and Computational Mechanics

Edited by F. Pfeiffer and P. Wriggers

Further volumes of this series found on our homepage: springer.com

Kinematics and Dynamics of Multibody Systems with Imperfect Joints

Models and Case Studies

Paulo Flores · Jorge Ambrósio · J.C. Pimenta Claro
Hamid M. Lankarani

With 127 Figures and 14 Tables

 Springer

P. Flores
Department of Mechanical Engineering
University of Minho
Campus Azurém
4800-058 Guimarães
Portugal

J. Ambrósio
Department of Mechanical Engineering
Instituto Superior Técnico
Technical University of Lisbon
Av. Rovisco Pais 1
1049-001 Lisboa
Portugal

J.C. Pimenta Claro
Department of Mechanical Engineering
University of Minho
Campus Azurém
4800-058 Guimarães
Portugal

Hamid M. Lankarani
Department of Mechanical Engineering
Wichita State University
Wichita, KS 67260-133
USA

ISBN   978-3-540-74359-0   e-ISBN   978-3-540-74361-3

ISSN   1613-7736

Library of Congress Control Number: 2007933833

© 2008 Springer-Verlag Berlin Heidelberg

Cover design: WMX Design GmbH, Heidelberg

Printed on acid-free paper

9 8 7 8 6 5 4 3 2 1 0

springer.com

Preface

The primary goal of this book is to present suitable methodologies for dynamic analysis of multibody mechanical systems with imperfect or real joints, that is, considering clearances, including their tribological characteristics and surface compliance properties. Two- and three-dimensional methodologies for imperfect kinematic joints with and without lubrication are presented. In the process, different contact-impact force models are revised in the face of their suitability to represent collision between the bodies connected by imperfect joints. The incorporation of friction forces, based on the Coulomb friction law, is also discussed together with an effective computational strategy. Further a general methodology which accounts for squeeze-film and wedge-film actions, including the cavitation effect, for modeling dynamically loaded journal–bearings is also presented.

The equations that govern the dynamical behavior of the general multibody mechanical systems incorporate the contact-impact forces due to collision between the bodies that constitute the imperfect joints, as well as the hydrodynamic forces owing to the lubrication effect. The Newton–Euler's equations of motion and the generalized Cartesian coordinates are used here. Elementary multibody mechanical systems are used to discuss the assumptions and procedures adopted. The main results obtained from this research work show that the effect of imperfect joints, namely joints with clearance, can have a predictable nonlinear dynamic response.

This book is written for academics, students and practitioners in mechanical engineering, design, researchers in the field and teaching staff. This book can also be used by students in the final year of MSc or in the beginning of PhD in mechanics and computation sciences.

Motivation and Contributions for this Book

Computer-aided multibody mechanical systems emerged over the last three decades as an important scientific part of Applied and Computational Mechanics, with significant applications in several branches of engineering. This has been made possible not only thanks to the impressive improvement of the computer hardware and software but also due to the development of robust and accurate computational tools, which in turn generated a demand for the analysis of more complex mechanical systems. Not long ago, the design of machines and components was based on trial and error and knowledgeable craftsmanship. Later, algebraic methods for analysis eliminated part of the limitations of the trial and error and led to documented methods used in the design of mechanical components. In today's industry, there is little room for error and a great need for optimized and cost-effective production of components and machines with high reliability and durability. Still, all areas of research require the construction of models and, therefore, the use of assumptions and approximations. The analysis of complex mechanical systems is an area where, in the past few decades, faster data processing has lead to an increased research effort. This includes kinematic and dynamic analyses, synthesis and optimization of the motion of mechanical systems. The main objective of this book is to contribute to the improvement of the methodologies dealing with the analysis of realistic mechanical systems, which include imperfect or real joints, i.e., joints in which the effects of clearance, misalignment, friction and lubrication are taken into account by employing the multibody systems formulation. In line with the multibody systems formulation, such systems can include several rigid and deformable bodies interconnected to each other by different types of kinematic joints and being acted upon by forces and moments. These systems are characterized by large displacements and rotations having, generally, a nonlinear behavior.

None of the real mechanical systems have perfect mechanical joints because of the functional tolerances required between adjacent moving segments. Due to either the loads carried by the mechanical devices or the misalignments that are required for their operation, real joints must be lubricated or must include bushings, generally made with metals and polymers. By using rubber bushings a conventional mechanical joint is transformed into a joint with clearance allowing for the mobility of the over-constrained system in which it is used. A good example of the application of these joints is vehicle dynamics in which the handling response or the vibration characteristics of land vehicles are greatly improved when these elements are used in the suspension system.

The mechanical joints of any industrial machine are obtained by allowing the relative motion between the components connected by them. Due to the manufacturing tolerances, wear or material deformation, these joints are imperfect and have clearances. These clearances modify the dynamic response of the system, justify the deviations between the numerical predictions and the experimental measurements and eventually lead to important deviations between the projected behavior of the mechanisms and their real outcome. The attenuation of the impact response and of the vibration characteristics in industrial machines is obtained by including, in their design, a selection of joint clearances.

The imperfect joints with direct contact between the parts involved generally use lubrication to minimize the energy dissipation. Therefore appropriate tribological models must be devised in the framework of their application in general multibody systems. The characterization of the normal contact forces in the nonlubricated joints is realized by using the continuous contact force model while their tangential forces are obtained by using appropriate friction force models. The joints, which have rubber bushings, can be described by a linear model that does not include coupling between radial and axial or bending loading or by a more advanced non-linear model for rubber bushings.

The general-purpose computational tools used for the design and analysis of mechanical systems have a wide number of mechanical systems modeling features that require the description of rigid or flexible bodies for which geometry, mass, center of mass, moment of inertia and other relevant properties are defined. The computational codes also provide a large library of kinematic joints that constrain relative degrees of freedom between connected bodies. The kinematic joints available in the commercial programs are represented as ideal joints, that is, there are no clearances or deformations in them. Thus modeling the dynamics of multibody mechanical systems with clearances and imperfections is a challenging issue in mechanical design and much work still remains to be done to achieve more advanced modeling tools.

This observation, coupled with the dearth of design guidelines in the field of mechanical systems with real or imperfect joints, and also the importance of such systems have motivated the work reported in this book.

Organization of the Book

An introduction and an overview of the book are provided in the first chapter. The remainder of this book consists of five more chapters, dealing with the different aspects involved in the study of imperfect joints in multibody mechanical systems.

In Chap. 2, the formulation of motion's equations of multi-rigid body systems is described. The generalized coordinates are the centroidal Cartesian coordinates and the system configuration is restrained by constraint equations. The dynamic formulation uses the Newton–Euler's equations of motion, which are augmented with the constraint equations that lead to a system of differential algebraic equations. Constraint violation stabilization methods and the coordinate partition method are also presented and discussed in this chapter. An elementary four-bar linkage is used as an application example to demonstrate the computational treatment of this type of systems.

Chapter 3 deals with contact-impact force models for both spherical- and cylindrical-shaped surface collisions in multibody mechanical systems. The inclusion of friction forces based on Coulomb's friction law is also presented and discussed for effective computational implementation.

Chapter 4 focuses on the modeling of planar clearance joints without lubricant, namely revolute and translational joints with clearance. Kinematic aspects of revolute and translational joints with clearance are also presented. Results for a basic slider–crank mechanism with a revolute clearance joint and a translational clearance joint are presented and used to discuss the assumptions and procedures adopted.

The fifth chapter covers the lubricated models for revolute clearance joints in multibody mechanical systems. The squeeze-film and wedge-film actions are considered as well as the cavitation effect. First, some techniques for modeling and evaluating the forces in lubricated revolute joints are presented and applied to a simple journal–bearing under a constant and unidirectional external load. Finally, a slider–crank mechanism with a lubricated revolute joint is considered as a numerical example. Both the effect of the clearance size and oil lubricant viscosity effects are studied.

Chapter 6 describes the modeling of spatial clearance joints, namely spherical and revolute joints. A quick review on the formulation of spatial multibody systems is presented. In addition, a simple and brief description of the perfect spherical and revolute joints is done. Three simple multibody systems are used as illustrative examples, namely, the spatial four-bar mechanism, the double pendulum and the slider–crank mechanism.

Acknowledgments

The road to this book has been long but rewarding. It has implied hard work, benefited from exceptional circumstances and fortunate opportunities, hesitated against minor difficulties, but eventually made its way receiving sometimes a stimulating recognition from the international scientific community. However, that journey has certainly not been the result of a solitary commitment. Many individuals and institutions have offered their support along the way, making it possible, visible and consequent. Certainly we will forever be in debt to them!

We are grateful to *Fundação para a Ciência e a Tecnologia and Fundo Comunitário Europeu FEDER* for sponsoring our research work, through project nr 38281, entitled 'Dynamic of Mechanical Systems with Clearances and Imperfections'. This work was also supported by PRODEP under the project 5.3/N/189.015/01. This support is thankfully acknowledged.

We are indebted to Professor Michel Fillon, University of Poitiers, for his stimulating comments and suggestions on the lubrication issues. We would also like to express our sincere gratitude to Professor Parviz Nikravesh, University of Arizona, for sharing with us some thoughts, ideas and computational codes on the dynamics of multibody systems. We are also grateful to Professor Werner Schiehlen, University of Stuttgart, for his feedback on the history of mechanics and multibody dynamics and to Professors Peter Eberhard, University of Stuttgart, and Cristoph Glocker, ETH Zurich, for their comments on the contact mechanics.

Finally, we would like to thank our families and our friends in different countries around the world for their never-ending encouragement, constant moral support and belief in our ability to do the work.

Contents

Nomenclature

All symbols in this book are described in the nomenclature, unless stated otherwise in the text.

General

Matrices are in boldface upper-case characters.
Vectors and column matrices are in boldface lower-case characters.
Scalars are in lightface characters.

Latin Symbols

Symbol	SI unit	Description
$\mathbf{A}$	–	Rotational transformation matrix
a	m	Length
c	m	Radial clearance
c_d	–	Dynamic correction coefficient
c_e	–	Coefficient of restitution
c_f	–	Coefficient of friction
D	Ns/m	Hysteresis damping coefficient
E	N/m^2	Young's modulus of elasticity
$\mathbf{e}$	–	Eccentricity vector
e	m	Eccentricity magnitude
$\dot{e}$	m/s	Eccentricity time derivative or radial velocity
e_0, e_1, e_2, e_3	–	Euler parameters
$\mathbf{f}$	–	Force vector
F_N	N	Normal contact force
F_T	N	Tangential force
$\mathbf{g}$	N, N m	Generalized force vector
$\mathbf{g}^{(c)}$	N, N m	Constraint reaction forces vector
H	m	Distance between the guide surfaces
h	m	Fluid film thickness
K	N/m$^{1.5}$	Generalized stiffness
L	m	Length

Symbol	SI unit	Description
$\mathbf{M}$	kg, kg m^2	System mass matrix
m	–	Number of kinematic independent constraints
n	–	Hertz's contact force exponent
nb	–	Number of bodies
nc	–	Number of Cartesian coordinates
$\mathbf{n}$	–	Normal unit vector
$\mathbf{p}$	–	Euler parameter vector
p	N/m^2	Fluid pressure
$\mathbf{q}$	–	Vector of generalized coordinates
$\dot{\mathbf{q}}$	–	Vector of velocities
$\ddot{\mathbf{q}}$	–	Vector that contains the state of accelerations
R	m	Radius
$\mathbf{r}$	–	Global position vector
$\mathbf{s}'$	–	Local position vector
t	s	Time
$\mathbf{t}$	–	Tangential unit vector
v_N	m/s	Normal velocity
v_T	m/s	Tangential velocity
W	m	Slider width
XY	–	2D global coordinate system
XYZ	–	3D global coordinate system
$\mathbf{y}$	m, m/s	System position and velocity
$\dot{\mathbf{y}}$	m/s, m/s^2	System velocity and acceleration

Greek Symbols

Symbol	SI unit	Description
$\mathbf{\Phi}$	–	Position constraints vectors
$\mathbf{\Phi_q}$	–	Jacobian matrix of constraints
Φ_{qt}	–	Time derivative of Jacobian matrix
$\dot{\mathbf{\Phi}}$	–	Constraint velocity equation
$\ddot{\mathbf{\Phi}}$	–	Constraint acceleration equation
α	–	Baumgarte stabilization coefficient
β	–	Baumgarte stabilization coefficient
χ	N s/m$^{2.5}$	Hysteretic damping factor
δ	m	Penetration depth
$\dot{\delta}$	m/s	Penetration velocity
$\dot{\delta}^{(-)}$	m/s	Initial impact velocity
ε	–	Eccentricity ratio
$\boldsymbol{\gamma}$	–	Right-hand side vector of acceleration equations
λ	–	Lagrange multipliers vector
μ	N s/m^2	Dynamic fluid viscosity
$\boldsymbol{\nu}$	–	Right-hand side vector of velocity equations
θ	rad	Angular position
ρ	kg/m^3	Material mass density
σ	m^2/N	Material parameter
ν	–	Poisson's ratio
ω	rad/s	Angular velocity
$\xi\eta$	–	2D body-fixed coordinate system
$\xi\eta\zeta$	–	3D body-fixed coordinate system

Subscripts

Symbol	Description
B	Bearing
i	Relative to body i
J	Journal
j	Relative to body j
n	Normal direction
q	Generalized coordinate
r	Radial direction
t	Tangential direction

Superscripts

Symbol	Description
0	Initial conditions
g	Ground
P	Generic point P
r	Revolute joint
s	Spherical joint
t	Translational joint
x	X-direction
y	Y-direction
z	Z-direction

Operators

Symbol	Description
$(\ )^T$	Matrix or vector transpose
$(')$	Components of a vector in a body-fixed coordinate system
$(\dot{\ })$	First derivative with respect to time
$(\ddot{\ })$	Second derivative with respect to time
$(.)$	Scalar or internal product
$(\times)$	Cross or external product
(∂)	Partial derivative
$(\sim)$	Skew-symmetric matrix or vector
Δ	Increment

Abbreviations

Symbol	Description
2D	two-dimensional,
3D	three-dimensional,
ADAMS	automatic dynamic analysis of mechanical systems,
ALF	augmented Lagrangian formulation,
BSM	Baumgarte stabilization method,

Symbol	Description
CAE	computer-aided engineering,
CPM	coordinate partitioning method,
DADS	dynamic analysis and design system,
DAE	differential algebraic equation,
DAP	dynamic analysis program,
DIM	direct integration method,
DIVPRK	double initial value problem Runge–Kutta,
DOF	degrees of freedom,
DRAM	dynamic response of articulated machinery,
EHL	elasto-hydrodynamic lubrication,
ESDU	Engineering Sciences Data Unit,
FEM	finite element method,
GC	geometric center,
IMP	integrated mechanism program,
IVPRK	initial value problem Runge–Kutta,
KAM	kinematic analysis method,
KAPCA	kinematic analysis program using complex algebra,
KINSYN	kinematic synthesis,
LINCAGES	linkage interactive computer analysis and graphically enhanced synthesis,
MADYMO	mathematical dynamical models,
MBS	MultiBody System,
ODE	ordinary differential equation.

Chapter 1
Introduction

Multibody dynamics can be understood as the study of systems of many bodies whose interactions are modeled by forces and kinematic constraints. In other words, a multibody system (MBS) can be defined as a collection of bodies acted upon by forces of different origins and interconnected to each other by different types of joints that constrain their motion. The forces applied to the system components may include those resulting from contact-impact, friction, gravity, joint constraints, external applications, the interaction with other systems such as fluids, tires or wheel–rail contact or due to mechanical elements such as springs, dampers and actuators. Kinematic constraint types may include revolute joints, translational joints, spherical joints and cylindrical joints, among others. The kinematic constraints may also be in the form of prescribed trajectories for given points of the system components or as driving constraints for a subsystem. The mechanical systems included under the definition of MBS comprise robots, heavy machinery, automobile suspensions and steering systems, machinery tools, animal bodies or satellites, among others (Wittenburg 1977, Nikravesh 1988). However, the range of systems that can be represented by MBS models is expanding and new exciting applications are being proposed everyday.

1.1 Multibody Systems and the Advent of Computers

Traditionally the kinematic and dynamic analyses of multibody systems was undertaken by assuming that the bodies were treated as rigid and without considering the physical properties of the joints (Reuleaux 1963, Martin 1982, Shigley and Uicker 1995). The kinematic solutions were obtained using the graphical method or analytical method for simple cases. The forces and moments on the bodies were then obtained based on the kinematic data. Since the traditional graphical and analytical methods are only available for simple cases of mechanisms and due to the broad usage of computers, numerical solutions for multibody systems were then developed (Chace 1967, Uicker 1969, Paul and Krajcinovic 1970).

It is well known that by the end of the eighteenth century the basic laws of motion for both translation and rotation were known. However, up to the middle of

P. Flores et al., *Kinematics and Dynamics of Multibody Systems with Imperfect Joints.*
© Springer 2008

the twentieth century little work effort was put on the dynamic analysis of systems of interconnected bodies undergoing general translation and rotation. In fact, the multibody system dynamics as it is known today began with the advent of computers.[1] The first and most important impetus for development of multibody dynamics was the rise of electronic digital computers, which finally broke down the massive computational barriers. It was not until the late 1960s that computational techniques found their way into the field of mechanical engineering.

In recent years greater emphasis has been placed on the design of high-speed, lightweight and precision systems. Generally these systems incorporate various types of driving, sensing and controlling devices working together to achieve specific performance requirements under different loading conditions. The design and performance analysis of such systems can be greatly enhanced through transient dynamic simulations, provided that all significant effects can be incorporated into the mathematical model. The need for a better design, in addition to the fact that many mechanical and structural systems operate in adverse environments, demanded the inclusion of many factors that have been ignored in the past. Systems such as engines, robotics, machine tools and space structures may operate at high speeds and in very high temperature environments. Neglecting deformation effects and clearances at the joints, for example, when these systems are analyzed, can lead to a mathematical model that poorly represents the actual system.

The research in the area of multibody dynamics has been motivated by a growing interest in the simulation and design of large-scale systems of interconnected bodies that undergo large displacements and rotations. The analysis and design of such systems require the simultaneous solution of hundreds or thousands of differential equations of motion, a task that could hardly be accomplished a few decades ago before the development of electronic computers. By the mid-1960s, attention turned to the possibility of constructing general-purpose multibody computer programs. With such programs, it is possible to simulate the motion of a very broad class of multibody systems. General-purpose programs are capable of generating and integrating the equations of motion, based on input data describing the way the bodies are interconnected, the mechanical and geometric properties of the bodies and the interaction between them, along with the system state at an initial time. This requires systematic techniques for formulating the equations of motion and numerical methods for solving them.

Broadly the rise of multibody systems dates back to the mid-1960s. Ever since, multibody dynamics has grown into one of the major fields of computational mechanics. In fact, the formalisms for computer-oriented generation of multibody systems have been developed to a high degree of maturity during the past decades. The specialized literature includes a number of books on the subject (Wittenburg 1977, Kane and Levinson 1985, Haug 1989, Nikravesh 1988, Shabana 1989, Huston 1990,

---

[1] In 1964 Gordon Moore—semiconductor engineer and co-founder of Intel in 1968—stated that the density of transistors in silicon based on micro-chips would double every year. Except in the late 1970s, where the doubling period increased to 18 months, this statement has been true for over three decades.

Schiehlen 1990, Amirouche 1992), where a broad variety of formulations and computational methods are discussed. More recently, a number of review papers of interest have been provided by Huston (1991), Schiehlen (1997), Shabana (1997) and Rahnejat (2000). Indeed, over the last years, the field of multibody systems has developed a new vigor inspired by new discoveries of formulation, new computational and experimental procedures and new applications. The modeling aspects in the various technological contexts and the interaction with other methodologies for computer-aided system design are fertile research topics in multibody dynamics (Ambrósio and Pereira 2003). Schiehlen (1997) stated that 'Challenging applications include biomechanics, chaos and nonlinearity, robotics and society and vehicle control. . . multibody system dynamics turn out to be a very lively and promising research subject' and such statement is being revised everyday by challenging applications and developments.

Over the last 40 years, several general-purpose computer codes that are capable of performing dynamic analysis of constrained mechanical systems have been developed (Chace 1967, Uicker 1969, Paul and Krajcinovic 1970). A program called KAPCA (cited by Shigley and Uicker 1995), standing for kinematic analysis program using complex algebra, written by students at the University of Wisconsin, has proved highly efficient as well as easy to use. This program had severe limitations such as the inability to perform force analysis of mechanisms. The first widely available general program for mechanism analysis was named KAM (kinematic analysis method) and was developed and distributed by IBM (Chace 1963, 1964). It included the capabilities for position, velocity, acceleration and force analysis of both planar and spatial systems. KINSYN (Rubel and Kaufman 1977) is a program available today which is intended primarily for kinematic synthesis. It addresses the synthesis of planar mechanism. This program was developed byKaufman (1978) at the Massachusetts Institute of Technology. The DRAM, an acronym for dynamic response of articulated machinery, is a generalized program for the kinematic and dynamic analyses of planar mechanisms that was developed by Chace (1978) at the University of Michigan and that is in the origin of the commercial code ADAMS. DRAM can be used to simulate even very complex planar mechanisms and provide position, velocity, acceleration and static force or dynamic force analysis. IMP (integrated mechanism program) was developed by Sheth and Uicker (1971) at the University of Wisconsin. This program can also be used to simulate either planar or spatial systems and provide kinematic, static and dynamic analyses (Erdman 1985, 1995, Crossley 1988). Erdman and Gustafson (1977) developed a computer code called LINCAGES (*l*inkage *i*nteractive *c*omputer *a*nalysis and *g*raphically *e*nhanced *s*ynthesis) to solve synthesis of mechanism problems.Orlandea et al. (1977) presented the first practical solution methodology for large multibody systems, based upon Lagrangian dynamics for constrained systems. This work was the basis of the computer program ADAMS, an acronym for automatic dynamic analysis of mechanical systems, the formulation and implementation of which incorporate stiff integration algorithms and sparse matrix techniques. Haug (1989) used the Cartesian position and orientation of the center of mass of each rigid body in the system to express the equations of motion. The constraints are determined from the kinematic pairs connecting adjacent bodies. The resulting dynamics algorithm has been implemented in the software package DADS, an acronym for dynamic analysis and design system (Haug et al. 1982). KAP (kinematic analysis program) and DAP (dynamic analysis program) are two FORTRAN codes developed by Nikravesh (1988) at the University of Arizona. These computer codes allow the kinematic and dynamic analyses of mechanical systems with rigid bodies. Garcia de Jálon and Bayo (1994) use the Cartesian coordinates of the joints connecting each body in the system, called natural coordinates, to derive the equations of motion. The constraint equations are derived from the relationship between adjacent kinematic pairs. Their formulation is used in the computer code COM-PAMM (Jiménez et al. 1990). Also in Europe there have been important multibody codes developed in the past 20 years. NEWEUL (Kreuzer 1979) is a software package that has been developed and maintained at the Stuttgart University since 1979 and is able to perform dynamic analysis of mechanical systems, such as vehicle dynamics, biomechanics and dynamics of mechanisms. NEWEUL software, based on the Newton–Euler formulation, generates the equation of motion of the multibody systems in symbolic form. MECANO (Géradin and Cardona 1989) is a software that allows the simulation of articulated systems and accounts for nonlinear flexibility and large displacements. SIMPACK (Rulka 1990) is a commercial multibody simulation code based on recursive dynamics that is able to describe and predict the motion of general mechanical systems, analyze vibrational behavior, calculate forces and moments and interface CAE environments. MADYMO, acronym for mathematical dynamical models, is an advanced software engineering tool developed by TNO (1998), which is basically used for occupant safety analysis. This software tool combines multibody and finite element methods and allows the design and optimization of vehicle structures, components and safety systems. Based on a recursive dynamics formalism the computer code RecurDyn (2006) is one of the most recent general-purpose commercial multibody codes capable of modeling a wide variety of complex multibody systems.

1.2 Multibody System's Equations of Motion

In a broad sense, the best-known methods to derive the equations of motion are Newton–Euler's method (Nikravesh 1988), Lagrange's method (Shabana 1989) and Kane's method (Kane and Levinson 1985). The Newton–Euler's methodology involves introducing a set of Lagrange's multipliers representing reaction forces of the joints which require an additional set of algebraic constraint equations. The size of the solution matrix becomes large, thereby occupying a substantial memory space for computation. However, this approach is relatively simple to implement. The Lagrange and Kane's methods select generalized coordinates to eliminate the explicit use of constraint equations from the formulation; therefore, a minimum number of equations are generated. The main drawback of these techniques concerns their difficulty to find generalized coordinates. Also their implementation is not as easy and straightforward as the Newton–Euler's method. For this reason, the Newton–Euler's method is adopted throughout this book to present the methodologies developed here.

The transient dynamic response of a constrained multibody system can be obtained by solving its governing equations of motion. In general, when not using a minimal set of coordinates, the motion of MBS is described by the so-called differential algebraic equations (DAE), i.e., a set of differential equations coupled with algebraic constraints (Nikravesh 1988, Haug 1989). The differential equations are of second order, and the algebraic equations describe the kinematic joints in the system. The solution of this type of equations and their integration in time introduces several numerical difficulties, namely the existence and uniqueness of solutions and instability for higher index systems (Garcia de Jálon and Bayo 1994). Specific numerical algorithms that enforce the stability of the solution are often required (Shampine and Gordon 1975, Gear 1981, Brenan et al. 1989). These algorithms use multi-time stepping procedures and often have the ability to deal with stiff systems. An alternative approach for the solution of the equations of motion is to transform the set of DAE in its underlying set of ordinary differential equations (ODE), which are solved by integration in time. It is well known that the substitution of the algebraic equations of the DAE system by their differential counterpart in the ODE system introduces mild instabilities and drift problems in the integration process, which can be attenuated using stabilization techniques such as Baumgarte stabilization method (BSM) (Baumgarte 1972) or the iterative augmented Lagrangian formulation (ALF) (Bayo et al. 1988). Other methods such as the K-U formulation, based on the application of the Moore Penrose generalized inverse/pseudo-inverse matrix to multibody systems, are also applied with success, increasing the efficiency and robustness of the formulation in the presence of redundant constraints (Neto and Ambrósio 2003), but they do not help in the stabilization of the constraint violations.

Due to its simplicity and computational implementation easiness, Cartesian coordinates are used in this work to formulate the equations of motion of the multibody systems. The kinematic constraints are added to the formulation by using the Lagrange multipliers technique. This simple, but powerful, approach allows for the analysis of complex mechanical systems. The equations of motion resulting from this formulation are composed of a set of differential algebraic equations (Nikravesh 1988). In a constrained multibody system, the equations of motion are solved by appending the constraint acceleration equations to the formulation. The resulting system of equations leads to a high-sparsity matrix. This fact can be exploited in order to improve the computational efficiency of the multibody programs. When solving the system of equations, the Gauss elimination scheme requires a high number of arithmetic operations that are proportional to the cube of the matrix dimension. With the use of a sparse matrix solver the number of arithmetic operations is greatly reduced, becoming almost proportional to the number of nonzero elements in the matrix of the system. The use of a sparse solver performs the Gauss elimination only on the nonzero elements of the matrix, reducing the number of fill-ins needed while showing good numerical stability. Such approach is used with success in rigid and flexible multibody systems by Gonçalves and Ambrósio (2000), Gonçalves (2002) and Pombo (2004).

The computational programs for multibody systems analysis are based on different approaches for automatic generation and integration of the equations of motion. The computational efficiency of these programs depends upon several factors, two of which are the choice of the type of coordinates and the methods used for the numerical solution of the different types of equations. The choice of coordinate type directly influences both the number of equations of motion and their order of nonlinearity. Furthermore, depending on the form of these equations, one method of numerical solution may be preferable to another in terms of efficiency and accuracy. Relative coordinates have the advantages of leading to the smallest set of equations possible. However, they present drawbacks such as the large complexity in their derivation, the high degree of nonlinearity of the equations of motion and their complex computer implementation. To overcome these disadvantages, Cartesian coordinates are used in most of the general-purpose computer codes, such as ADAMS (Ryan 1990), which is broadly used in industry. The formulation with Cartesian coordinates yields a large number of equations but the order of nonlinearity is lower than that of the relative, or point, coordinates.

Although in this work all methodologies presented use Cartesian coordinates it can be argued that depending on the application different types of coordinates present relative advantages and drawbacks. Also, depending on how the formulations based on a particular type of coordinates are implemented computationally, some of the drawbacks on the use of such coordinates may be limited or eliminated. A review on advantages and disadvantages of different formulations can be found in Nikravesh (1988).

1.3 Kinematic and Dynamic Analyses of Multibody Systems

In dealing with the study of multibody system motion, two different types of analysis can be performed, namely, the kinematic analysis and the dynamic analysis. The kinematic analysis consists in the study of the system's motion independently of the forces that produce it, involving the determination of position, velocity and acceleration of the system components. In kinematic analysis, only the interaction between the geometry and the motions of the system is analyzed and obtained. Since the interaction between the forces and the system's motion is not considered, the motion of the system needs to be specified to some extent, that is, the kinematic characteristics of some driving elements need to be prescribed, while the kinematic motion characteristics of the remaining elements are obtained using the kinematic constraint equations, which describe the topology of the system.

The dynamic analysis of multibody systems aims at understanding the relationship between the motion of the system parts and the causation of the motion, including external applied forces and moments. The motion of the system is, in general, not prescribed, its calculation being one of the principal objectives of the analysis. The dynamic analysis also provides a way to estimate external forces that depend on the relative position between the system components, such as the forces exerted by springs, dampers and actuators. Furthermore it is also possible to estimate the external forces that are developed as a consequence of the interaction between the system components and the surrounding environment, such as contact-impact and friction forces. The internal reaction forces and moments generated at the kinematic joints are also obtained in the course of the dynamic analysis. These reaction forces and moments prevent the occurrence of the relative motions, in prescribed directions, between the bodies connected via kinematic joints. The fundamental aspects related to the kinematic and dynamic analyses of multibody systems are discussed in detail in the forthcoming chapters.

1.4 Multibody Systems with Imperfect Joints

Multibody systems are made of several components, which can be divided into two major groups, namely links, that is, bodies with a convenient geometry, and kinematic joints, which introduce some restrictions on the relative motion of the various bodies of the system. Usually the bodies are modeled as rigid or deformable bodies, while joints are represented by a set of kinematic constraints. The functionality of a kinematic joint relies upon the relative motion allowed between the connected components. This fact implies the existence of a gap, that is, a clearance between the mating parts, and thus surface contact, shock transmission and the development of different regimes of friction and wear. No matter how small that clearance is, it can lead to vibration and fatigue phenomena, lack of precision or even random overall behavior. Therefore it is quite important to quantify the effects of both clearance joints and bodies flexibility on the global system response in order to define the minimum level of suitable tolerances that allow mechanical systems to achieve required performances.

Mechanical systems with rigid and flexible bodies and with nonideal joints have been treated in the past. Studies have considered joint compliance and friction, but without clearances (Haug et al. 1986). Methods for modeling joint connections and external impacts using restitution coefficient and momentum have also been proposed in the literature (Khulief and Shabana 1986, 1987). While such methods offer the advantage of relatively low computational effort over techniques that explicitly model the joint impact, they do not provide the values of these forces, which are important from the design point of view. In fact, measurements of the influence that joint clearances have on the kinematic and dynamic performances of a mechanism provide useful criteria for judging the suitability of the mechanism to perform a given task.

In general, in the dynamic analysis of multibody mechanical systems it is assumed that the kinematic joints are ideal or perfect, that is, clearance, local elastic/plastic deformations, wear and lubrication effects are neglected. However, in a real mechanical joint a gap is always present. Such clearance is necessary not only to allow the relative motion between the connected bodies but also to permit the assemblage of components. For instance, in a journal–bearing joint, such as the one shown in Fig. 1.1, there is a radial clearance allowing for the relative motion between the journal and the bearing. The presence of such joint clearance leads to degradation of the performance of mechanical systems in virtue of the impact forces that take place. No matter how small the clearance is, it can lead to vibration and fatigue phenomena, premature failure and lack of precision or even random overall behavior.

The degradation of the performance of mechanical systems with clearance joints has been recognized for a number of years (Goodman 1963). It is known that the forces within joints with realistic amounts of clearance, due to clearance impacts, are much higher than would be calculated if the joint clearances were neglected (Flores and Ambrósio 2004). High joint impact forces can also cause early failure. However, the total elimination of the clearance in the connections aiming at eliminating joint impacts is, usually, either expensive or impossible.

The subject of the representation of imperfect joints has attracted the attention of a large number of researchers that produced a number of theoretical and experimental works devoted to the dynamic simulation of mechanical systems with joint clearances (Dubowsky and Freudenstein 1971a, b, Bahgat et al. 1979, Grant and Fawcett 1979, Haines 1985, Soong and Thompson 1990, Deck and Dubowsky 1994, Ravn 1998, Claro and Fernandes 2002). Some of these works focus on the planar systems in which only one kinematic joint is modeled as a clearance joint. Dubowsky and Freudenstein (1971a, b) formulated an impact pair model to predict the dynamic response of an elastic mechanical joint with clearance. In their model, springs and dashpots were arranged as Kelvin–Voigt model. This work was subsequently extended by Dubowsky and his co-workers (Dubowsky 1974a, Dubowsky and Gardner 1977, Dubowsky and Moening 1978). Dubowsky (1974b) showed how

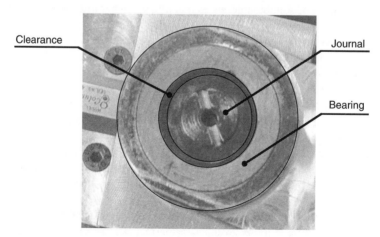

Fig. 1.1 Actual revolute joint with clearance

clearances can interact dynamically with machine control systems to destabilize and produce undesirable limit cycle behaviors.

Wilson and Fawcett (1974) assumed that clearance exists in the sliding bearing in a slider–crank mechanism. A theoretical investigation of the effects of parameters, such as the geometry, speed and mass distribution of the mechanism, upon the transverse motion of the slider was reported by these researchers. They derived the equations of motion for the different scenarios of the slider motion inside the guide.

Earles and Wu (1973) employed a modified Lagrange's equation approach in which constraints were incorporated using Lagrange multipliers in order to predict the behavior of rigid mechanism with clearance in a journal–bearing. The clearance in the journal–bearing was modeled by a massless imaginary link, but the simulation was restricted to the range of motion that starts when the contact between the journal and the bearing is terminated. Later they investigated the prediction of contact loss in a bearing of a linkage mechanism (Earles and Wu 1975, Wu and Earles 1977).

Mahrus (1974) conducted an experimental investigation into journal–bearing performance. A test machine was developed on which steady or varying unidirectional or full two-component dynamic load can be applied to the test journal–bearing. The corresponding journal center path is measured simultaneously with the load to show the effect of the load diagram on hydrodynamic lubrication.

Grant and Fawcett (1979) proposed a method to prevent contact loss between the journal and the bearing. Their experimental results verified the approach for a limited class of systems, but could not overcome the lack of universality of the method (Haines 1980a). In a theoretical study, Townsend and Mansour (1975) modeled a four-bar crank-rocker mechanism with clearance as two sets of compound pendulums. They ignored the motion in contact mode entirely and instead a close succession of small pseudo-impacts was assumed for the simulation. Subsequently Miedema and Mansour (1976) extended their previous two-mode model, for the free flight and impact modes, to a three-mode model in which a following mode was proposed. In their numerical simulations, the following mode was always assumed to occur immediately after the impact mode. However, this is frequently not observed in practice.

Haines (1980b) derived equations of motion that describe the contributions at a revolute joint with clearance but with no lubrication present. Also Haines (1985) carried out an experimental investigation on the dynamic behavior of revolute joints with varying degrees of clearance. Under static loads, the deflection associated with contact elasticity in the dry journal–bearing is found to be much larger and less linear than predicted.

Bengisu et al. (1986) developed a separation parameter for a four-bar linkage which was based on a zero-clearance analysis. The theoretical results were compared with the experimental results and showed good qualitative agreement. They also predicted contact loss in a mechanism with multiple clearance joints.

Norton et al. (1994) discussed the bearing forces as a function of mechanical stiffness and the vibration isolation in an experimental four-bar mechanism. The test rig was instrumented with piezoelectric accelerometers and force transducers. As expected, the clearances in the physical model created significantly larger dynamic accelerations, torques and forces on the bearings than was predicted theoretically by the rigid dynamic model.

Roger and Andrews (1977) developed mathematical models for the journal–bearing elements, which take into account the effect of clearance, surface compliance and lubrication. However, their lubrication model only accounts for the squeeze-film effect. Later Liu and Lin (1990) extended Roger and Andrews' work to include both squeeze-film and wedge-film actions. Schwab (2002), based on the work of Moes et al. (1986), applied the impedance method to model lubricated revolute joints in a slider–crank mechanism. Flores et al. (2003) proposed a hybrid model for revolute clearance joints in which the dry contact and the pure squeeze-film effect are combined. In a later work, Flores et al. (2006), based on the Pinkus and Sternlicht (1961) formulation for dynamically loaded journal–bearings, presented a model for dynamic analysis of a planar slider–crank mechanism with a lubricated revolute joint. Alshaer et al. (2001) analyzed and compared the dynamic response of a slider–crank mechanism in which a revolute joint was modeled as frictionless dry contact and lubricated joint as well.

Feng et al. (2002) developed an optimization method to control the change of inertia forces by optimizing the mass distribution of moving links in planar linkages with clearances at joints. Innocenti (2002) presented a method for analysis of spatial structures with revolute clearance joints. García Orden et al. (2003) presented a methodology for the study of typical smooth joint clearances in multibody systems. This approach takes advantage of the analytical definition of the material surfaces defining the clearance, resulting in a formulation where the gap does not play a central role, as it happens in standard contact models.

Other researchers also included the influence of the flexibility of bodies in the dynamic performance of multibody systems besides the existence of gaps in the joints (Winfrey et al. 1973, Dubowsky et al. 1991, Bauchau and Rodriguez 2002). Soong (1988) presented an analytical and experimental investigation of the elastodynamic response characteristics of a planar linkage with bearing clearances. Dubowsky and Moening (1978) demonstrated a reduction in the impact force level by introducing the flexibility of bodies. They also observed a significant reduction of the acoustic noise produced by the impacts when the system incorporates flexible bodies. Kakizaki et al. (1993) presented a model for spatial dynamics of robotic manipulators with flexible links and joint clearances, where the effect of the clearance is taken to control the robotic system.

The dynamic behavior of the multibody systems with clearance joints is very sensitive to small changes of parameters, namely the clearance size and the friction coefficient. In fact, the system response can change from periodic to chaotic with a very small variation of the initial parameters. Seneviratne and Earles (1992) used a massless link model to study the behavior of a four-bar mechanism with a clearance joint. Farahanchi and Shaw (1994) studied the dynamic performance of the slider–crank mechanism with a slider clearance. This research showed that both periodic and chaotic behaviors depend on the values of friction coefficient and crank velocity. Rhee and Akay (1996) used a discontinuous contact force model to analyze a four-bar mechanism with one revolute clearance joint. They showed that, depending on the friction coefficient and clearance size, the system can exhibit periodic or chaotic behavior. More recently, Ravn (1998) and Flores (2005) demonstrated, through the use of Poincaré maps, that the dynamic response of mechanical systems with clearance joints can be periodic in some situations, but also chaotic for other conditions.

1.5 Contact Analysis in Multibody Systems

In general, the motion characteristics of a multibody system are significantly affected by contact-impact phenomena. According to Gilardi and Sharf (2002), impact is a complex physical phenomenon for which the main characteristics are a very short duration, high force levels, rapid energy dissipation and large changes in the velocities of bodies. Inherently contact implies a continuous process which takes place over a finite time.

The problem of collision between rigid bodies is quite old and it was analyzed originally by Newton (1686). Whittaker (1904) extended Newton's work to account for friction phenomenon. Up to 1904, all of the contact problems used the theory developed by Newton, the inclusion of friction in the model being the major difficulty, as pointed out by Painlevé (1905). However, Kane (1984) pointed out, in a newsletter with limited circulation, an apparent paradox on the application of the Newton's theory with Coulomb's friction to a problem of a collision in a double pendulum leading to an increase of energy in the system. Naturally the question that arises is: what was wrong? Keller (1986) presented a solution to Kane's paradox, but the solution was not easy to generalize. Keller's solution resulted in widespread interest, which led to the development of relevant work and even to the publication of relevant textbooks totally devoted to this issue such as Brach (1991), Pfeiffer and Glocker (1996), Johnson (1999), Stronge (2000), among others.

Brach (1991) presented an algebraic solution scheme, revising Newton's theory and introducing impulse ratios, to describe the behavior in tangential directions. He defined the tangential impulse as a constant fraction of the normal impulse. This fraction is equivalent to friction coefficient in many cases. However, Brach's solution did not give clear solutions to the problem when the reverse sliding was considered during the collision. Smith (1991) proposed another purely algebraic approach to the problem using the Newton's theory for the coefficient of restitution. The impulsive ratio is determined using as velocity the average value of different slipping velocities. Wang and Mason (1992), based on the Routh's (1905) technique, compared the coefficients of restitution given by Newton and Poisson. They also identified the impact conditions under which Newton's and Poisson's models give the same solution. Stronge (1991) demonstrated the energy inconsistencies in some solutions obtained with Poisson's model when the coefficient of restitution is assumed to be independent of the coefficient of friction. Furthermore Stronge proposed to define the coefficient of restitution as the square root of the ratio of the elastic strain energy released during the restitution to the energy absorbed by deformation during compression.

Hurmuzlu and Marghitu (1994) studied the contact problem in multibody systems, where a planar rigid-body kinematic chain undergoes an external impact and an arbitrary number of internal impacts. Based on Keller's (1986) work, they developed a differential–integral approach and used different models for coefficient of friction. Han and Gilmore (1993) proposed a similar approach, using an algebraic formulation of the equations of motion, the Poisson's model of restitution and the Coulomb's law to define the tangential motion. Different conditions that characterize the motion (slipping, sticking and reverse sliding) are detected by analyzing velocities and accelerations at the contact points. Han and Gilmore (1993) confirmed their simulation results with experiments for two-body and three-body impacts. Haug et al. (1986) directly solved the differential equations of motion by using the Lagrange multiplier technique. Newton's model was used for impact while the Coulomb's law was used for friction.

The analysis of contact between two bodies can be extended to the analysis of impact in a multibody mechanical system. Whenever a link in a multibody system experiences a hard stop, contact forces of complex nature act on the links and the corresponding impulse is transmitted throughout the system. In the case of worn joints with clearances, the link moves freely inside the clearance zone, because it is not constrained, until it impacts onto the joint and the link experiences a hard stop (Flores and Ambrósio 2004). The methodology used to describe multibody mechanical systems with hard stops and joint clearances is based on the development of different contact force models that include energy dissipation, with which a continuous analysis of the system undergoing internal impact is performed.

The impact phenomenon is characterized by abrupt changes in the values of system variables, most commonly discontinuities in the system velocities. Other effects directly related to the impact phenomena are those of vibration propagation through the system, local elastic/plastic deformations at the contact zone and frictional energy dissipation. Impact is a prominent phenomenon in many mechanical systems such as mechanisms with intermittent motion and mechanisms with clearance joints (Lee and Wang 1983, Khulief and Shabana 1986, 1987). Therefore, in order to correctly simulate and design these kinds of mechanical systems adequately, some appropriate contact-impact force model must be adopted. Furthermore, during impact, a multibody system presents discontinuities in geometry and some material properties can be modified by the impact itself.

In a broad sense, the different methods to solve the impact problem in multibody mechanical systems are continuous and discontinuous approaches (Lankarani and Nikravesh 1990). Within the continuous approach, the methods commonly used are the continuous force model, which is in fact a penalty method, and the unilateral constraint methodology, based on complementary approaches (Pfeiffer and Glocker 1996). The continuous contact force model represents the forces arising from collisions and assumes that the forces and deformations vary in a continuous manner. In this method, when contact between the bodies is detected, a contact force perpendicular to the plane of collision is applied. This force is typically applied as a spring–damper element, which can be linear, the Kelvin–Voigt model (Timoshenko and Goodier 1970), or nonlinear, the Hunt and Crossley model (Hunt and Crossley 1975). Figure 1.2 schematically shows these two contact force models. For long impact durations this method is effective and accurate in so far as the instantaneous contact forces are introduced into the system's equations of motion. In the second continuous approach, when contact is detected a kinematic constraint is introduced in the system equations. Such a constraint is maintained while the reaction forces are compressive and removed when the impacting bodies rebound from contact (Ambrósio 2000).

As far as the second approach is concerned, the discontinuous method assumes that the impact occurs instantaneously and the integration of the equations of motion is halted at the time of impact. Then a momentum balance is performed to calculate the post-impact velocities of the bodies involved in the contact and all those included in the same kinematic chain. The integration is then resumed with the updated velocities until the next impact occurs. In the discontinuous method, the dynamic analysis of the system is divided into two intervals, before and after impact. The restitution coefficient is employed to quantify the dissipation energy during the impact. The restitution coefficient only relates the velocity just after separation and the velocity just before contact, and ignores what happens in between. The discontinuous method is relatively efficient. However, the unknown duration of the impact limits its application because if the duration of impact is large enough, the system configuration changes significantly (Lankarani 1988). Hence the assumption of instantaneity of impact is no longer valid and the discontinuous analysis cannot be adopted. This method, commonly referred to as piecewise analysis, has been used for solving intermittent motion problems (Khulief and Shabana 1986, 1987).

Broadly an impact may be considered to occur in two phases: the compression or loading phase and the restitution or unloading phase. During the compression phase, the two bodies deform in the normal direction to the impact surface, and the relative velocity of the contact points/surfaces on the two bodies in that direction is reduced

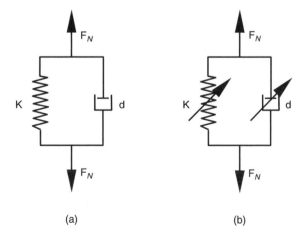

Fig. 1.2 Contact force models; F_N is the normal contact force, K is the stiffness constant and d is the damping coefficient.
(**a**) Linear viscous-elastic Kelvin–Voigt force model,
(**b**) nonlinear viscous-elastic Hunt and Crossley force model

to zero. The end of the compression phase is referred to as the instant of maximum compression or maximum approach. The restitution phase starts at this point and ends when the two bodies separate (Brach 1991). The restitution coefficient reflects the type of collision, for a fully elastic contact $c_e = 1$, while for a fully plastic contact $c_e = 0$. The most general and predominant type of collision is the oblique eccentric collision, which involves both relative normal velocity and relative tangential velocity (Maw et al. 1975, Zukas et al. 1982). Figure 1.3 shows the oblique eccentric collision between two bodies. A more detailed analysis on the contact-impact force models for multibody systems will be presented in Chap. 3.

The contact between two bodies can also be classified as conformal and nonconformal (ESDU 78035 1978, Johnson 1999). A contact is said to be conformal if the surfaces of the two bodies fit exactly or even closely together without deformation. Bodies which have dissimilar profiles are said to be nonconformal. When brought into contact without deformation they will first touch at a point (point contact) or along a line (line contact). Figure 1.4 illustrates examples of conformal and nonconformal contacts. It should be highlighted that the radius of curvature is, by definition, positive for convex surfaces and negative for concave surfaces.

The compliant continuous contact force models, commonly referred to as penalty methods, have been gaining significant importance in the context of multibody systems with contacts thanks to their computational simplicity and efficiency. In these models, the contact force is expressed as a continuous function of the penetration of relative bodies. However, one of the main drawbacks associated with these force models deals with the difficulty to choose the contact parameters such as the equivalent stiffness or the degree of nonlinearity of the penetration, especially for complex contact scenarios. Thus the complementary formulations for contact modeling in multibody systems have attracted the attention of many researchers (Rockafellar 1972, Moreau 1988, Pang et al. 1992, Pfeiffer 1999, Bauchau et al. 2001, Trinkle 2003). Assuming that the contacting bodies are truly rigid, as opposed to locally deformable or penetrable as in the penalty approaches, the complementary formulations resolve the contact dynamics problem by using the unilateral constraints to compute contact impulses or forces to prevent penetration from occurring. Thus at the core of the complementary approach is an explicit formulation of the unilateral constraints between the contacting rigid bodies (Kilmister and Reeve 1966, Rosenberg 1977, Brogliato et al. 2002). In the approach, contacts are

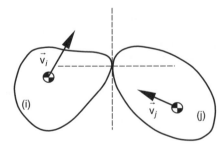

Fig. 1.3 Oblique eccentric collision between two bodies

Fig. 1.4 Conformal and
nonconformal contacts:
(**a**) sphere on sphere,
nonconformal contact,
(**b**) cylinder on cylinder,
conformal contact

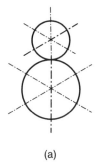

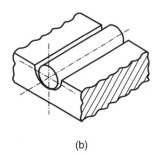

(a)   (b)

intermittent, that is, an active constraint can become passive or inactive, while an inactive constraint can become active. In the event of a passive constraint becoming active, the change is generally accomplished by an impact which is characterized by impulsive forces. Such an event changes the behavior of the system, the number of kinematic constraints and, therefore, the number of degrees of freedom of the system is changed. Consequently the governing system of differential equations of motions is changed too.

Lötstedt (1982) presented a formulation for the contact force developed between rigid bodies when subjected to unilateral constraints and constructed a unique solution by using the linear complementary theory of mathematical programming. Analysis of the bounds on contact velocity and existence of the contact force in the linear complementary formulation was also presented. Baraff (1989) provided the detail of the complimentarity conditions in acceleration–force form for the contact between two polyhedral bodies and proposed to employ heuristic solution methods to calculate the forces between the bodies in noncolliding contact. Pfeiffer and Glocker (1996) presented a formulation of the impact system dynamics with unilateral constraints at acceleration–force level and employed the impact rule formulated to convert the two-dimensional frictional impact problem to a linear complementary problem. The extension of the treatment to three-dimensional contact was developed and presented by Glocker (1999), in which both approximate polyhedral and conic friction laws were used to produce linear and nonlinear complementary formulations, respectively. Later Glocker (2001) presented a comprehensive revision of the contact dynamics fundamentals along with a number of practical contact examples. Pfeiffer (1999) presented a brief development of the contact force calculation for nonchanging contact configuration. Trinkle (2003) proposed a time-stepping algorithm incorporating Coulomb friction and inelastic impacts and shocks in velocity–impulse form.

Certainly the advances in contact mechanics and the extensive work published in this field allow finding advantages and drawbacks on the use of the different continuous and discontinuous approaches. In this work, the contact forces are represented using the continuous force approach due to its simplicity and its easy understanding. Although not discussed any further, unilateral constraints could also have been chosen to handle the contact. In any case, the contact detection and the methodologies to deal with the tribological effects remain the same regardless of the choice of methodology selected to handle the contact forces being based on unilateral constraints or on a penalty method.

References

Alshaer BJ, Lankarani HM, Shivaswamy S (2001) Dynamics of a multibody mechanical system with lubricated long journal bearings. Proceedings of DETC'01, ASME 2001 design engineering technical conferences and computers and information in engineering conference, Pittsburg, PA, September 9–12, 6pp.

Ambrósio J (2000) Rigid and flexible multibody dynamics tools for the simulation of systems subjected to contact and impact conditions. European Journal of Solids A/Solids 19: S23–44.

Ambrósio J, Pereira MS (2003) Desenvolvimentos recentes no cálculo automático de sistemas mecânicos: teoria a aplicação. Proceedings of VI Congresso Ibero-Americano de Engenharia Mecânica—CIBEM6, Departamento de Engenharia Mecânica, Universidade de Coimbra, Coimbra, edited by AM Dias, Vol. I, pp. 1–20.

Amirouche FML (1992) Computational methods for multibody dynamics. Prentice Hall, Englewood Cliffs, NJ.

Bahgat BM, Osman MOM, Sankar TS (1979) On the effect of bearing clearances in the dynamic analysis of planar mechanisms. Journal of Mechanical Engineering Science 21(6):429–437.

Baraff D (1989) Curves surfaces and coherence for non-penetrating rigid body simulation. Computer Graphics 24:19–28.

Bauchau OA, Rodriguez J (2002) Modelling of joints with clearance in flexible multibody systems. International Journal of Solids and Structures 39:41–63.

Bauchau OA, Rodriguez J, Bottasso CJ (2001) Modelling of unilateral contact conditions with application to aerospace systems involving blacklash, freeplay and friction. Mechanics Research Communication 28:571–599.

Baumgarte J (1972) Stabilization of constraints and integrals of motion in dynamical systems. Computer Methods in Applied Mechanics and Engineering 1:1–16.

Bayo E, Garcia de Jálon J, Serna A (1988) A modified Lagrangian formulation for the dynamic analysis of constrained mechanical systems. Computer Methods in Applied Mechanics and Engineering 71:183–195.

Bengisu MT, Hidayetoglu T, Akay A (1986) A theoretical and experimental investigation of contact loss in the clearances of a four-bar mechanism. Journal of Mechanisms, Transmissions, and Automation in Design 108:237–244.

Brach RM (1991) Mechanical impact dynamics, rigid body collisions. Wiley, New York.

Brenan KE, Campbell SL, Petzold LR (1989) The numerical solution of initial value problems in differential-algebraic equations. North-Holland, New York.

Brogliato B, Ten Dam AA, Paoli L, Genot F, Abadie M (2002) Numerical simulations of finite dimensional multibody nonsmooth mechanical systems. Applied Mechanics 55:107–150.

Chace MA (1963) Vector analysis of linkages. Journal of Engineering for Industry, Series B 55(3):289–297.

Chace MA (1964) Development and application of vector mathematics for kinematic analysis of three dimensional mechanisms. Ph.D. Dissertation, University of Michigan, Michigan.

Chace MA (1967) Analysis of the time-dependence of multi-freedom mechanical systems in relative coordinates. Journal of Engineering for Industry 89:119–125.

Chace MA (1978) Using DRAM and ADAMS programs to simulate machinery vehicles. Agricultural Engineering 59(11):16–18.

Claro JCP, Fernandes JPF (2002) Influência da modelização das juntas na análise do desempenho de um mecanismo. 8as Jornadas Portuguesas de Tribologia, Universidade de Aveiro, Aveiro, May 8–9, edited by J Grácio, P Davim, QH Fan and N Ali, pp. 215–219.

Crossley FE (1988) Recollections from forty years of teaching mechanisms. Journal of Mechanisms, Transmissions, and Automation in Design 110:232–242.

Deck JF, Dubowsky S (1994) On the limitations of predictions of the dynamic response of machines with clearance connections. Journal of Mechanical Design 116:833–841.

Dubowsky S (1974a) On predicting the dynamic effects of clearances in planar mechanisms. Journal of Engineering for Industry, Series B 96(1):317–323.

Dubowsky S (1974b) On predicting the dynamic effects of clearances in one-dimensional closed loop systems. Journal of Engineering for Industry, Series B 96(1): 324–329.

Dubowsky S, Freudenstein F (1971a) Dynamic analysis of mechanical systems with clearances, part 1: formulation of dynamic model. Journal of Engineering for Industry, Series B 93(1):305–309.

Dubowsky S, Freudenstein F (1971b) Dynamic analysis of mechanical systems with clearances, part 2: dynamic response. Journal of Engineering for Industry, Series B 93(1):310–316.

Dubowsky S, Gardner TN (1977) Design and analysis of multilink flexible mechanism with multiple clearance connections. Journal of Engineering for Industry, Series B 99(1):88–96.

Dubowsky S, Moening MF (1978) An experimental and analytical study of impact forces in elastic mechanical systems with clearances. Mechanism and Machine Theory 13:451–465.

Dubowsky S, Gu PY, Deck JF (1991) The dynamic analysis of flexibility in mobile robotic manipulator systems. Proceedings of VIII world congress on the theory of machines and mechanisms, Prague, Czechoslovakia, August 26–31, 12pp.

Earles SWE, Wu CLS (1973) Motion analysis of a rigid link mechanism with clearance at a bearing using lagrangian mechanics and digital computation. Mechanisms 83–89.

Earles SWE, Wu CLS (1975) Predicting the occurrence of contact loss and impact at the bearing from a zero-clearance analysis. Proceedings of IFToMM fourth world congress on the theory of machines and mechanisms, Newcastle Upon Tyne, England, pp. 1013–1018.

Erdman AG (1985) Computer-aided design of mechanisms: 1984 and beyond. Mechanism and Machine Theory 20(4):245–249.

Erdman AG (1995) Computer-aided mechanism design: now and the future—50th anniversary of the Design Engineering Division Combined Issue. Journal of Mechanical Design 117(B):93–100.

Erdman AG, Gustafson, JE (1977) LINCAGES: Linkage INteractive Computer Analysis and Graphically Enhanced Synthesis package. ASME paper No. 77 DTC-5.

ESDU 78035 Tribology Series (1978) Contact phenomena. I: stresses, deflections and contact dimensions for normally loaded unlubricated elastic components. Engineering Sciences Data Unit, London, England.

Farahanchi F, Shaw SW (1994) Chaotic and periodic dynamics of a slider crank mechanism with slider clearance. Journal of Sound and Vibration 177(3):307–324.

Feng B, Morita N, Torii T (2002) A new optimization method for dynamic design of planar linkage with clearances at joints—optimizing the mass distribution of links to reduce the change of joint forces. Journal of Mechanical Design 124:68–73.

Flores P (2005) Dynamic analysis of mechanical systems with imperfect kinematic joints. Ph.D. Dissertation, Universidade do Minho, Guimarães, Portugal.

Flores P, Ambrósio J (2004) Revolute joints with clearance in multibody systems. Computers and Structures, Special Issue: Computational Mechanics in Portugal 82(17–18):1359–1369.

Flores P, Ambrósio J, Claro JP (2003) Dynamic analysis for planar multibody mechanical systems with real joints. Proceedings of ECCOMAS multibody dynamics 2003, International conference on advances in computational multibody dynamics 2003, Lisbon, Portugal, July 1–4, edited by JAC Ambrósio, 26pp.

Flores P, Ambrósio J, Claro JCP, Lankarani HM, Koshy CS (2006) A study on dynamics of mechanical systems including joints with clearance and lubrication. Mechanism and Machine Theory 41(3):247–261.

Garcia de Jálon J, Bayo E (1994) Kinematic and dynamic simulations of multibody systems. Springer, Berlin Heidelberg New York.

García Orden JC, Romero I, Goicolea JM (2003) Analysis of joint clearances in multibody systems. Proceedings of ECCOMAS multibody dynamics 2003, International conference on advances in computational multibody dynamics 2003, Lisbon, Portugal, July 1–4, edited by JAC Ambrósio, 9pp.

Gear CW (1981) Numerical solution of differential-algebraic equations. IEEE Transactions on Circuit Theory CT-18:89–95.

Géradin M, Cardona A (1989) Kinematics and dynamics of rigid and flexible mechanisms using finite elements and quaternion algebra. Computational Mechanics 4:115–135.

Gilardi G, Sharf I (2002) Literature survey of contact dynamics modeling. Mechanism and Machine Theory 37:1213–1239.

Glocker C (1999) Formulation of spatial contact situations in rigid multibody systems. Computer Methods in Applied Mechanies and Engineering 177:199–214.

Glocker C (2001) Set-valued force laws—dynamics of non-smooth systems. Springer, Berlin Heidelberg New York.

Gonçalves J (2002) Rigid and flexible multibody systems optimization for vehicle dynamics. Ph.D. Dissertation, Universidade Técnica de Lisboa, Instituto Superior Técnico, Lisbon, Portugal.

Gonçalves J, Ambrósio J (2000) Advanced modeling of flexible multibody systems using virtual bodies. NATO ARW on computational aspects of nonlinear structural systems with large rigid body motion, edited by J. Ambrósio and M. Kleiber. IOS Press, The Netherlands.

Goodman TP (1963) Dynamic effects of backlash. Machine Design 35:150–157.

Grant SJ, Fawcett JN (1979) Effects of clearance at the coupler-rocker bearing of a 4-bar linkage. Mechanism and Machine Theory 14:99–110.

Haines RS (1980a) Survey: 2-dimensional motion and impact at revolute joints. Mechanism and Machine Theory 15:361–370.

Haines RS (1980b) A theory of contact loss at resolute joints with clearance. Proceedings of Institution of Mechanical Engineers, Journal of Mechanical Engineering Science 22(3):129–136.

Haines RS (1985) An experimental investigation into the dynamic behaviour of revolute joints with varying degrees of clearance. Mechanism and Machine Theory 20(3):221–231.

Han I, Gilmore BJ (1993) Multi body impact motion with friction analysis, simulation, and validation. Journal of Mechanical Design 115:412–422.

Haug EJ (1989) Computer-aided kinematics and dynamics of mechanical systems. Volume I: basic methods. Allyn and Bacon, Boston, MA.

Haug EJ, Wehage RA, Barman NC (1982) Dynamic analysis and design of constrained mechanical systems. Journal of Mechanical Design 104:778–784.

Haug EJ, Wu SC, Yang SM (1986) Dynamics of mechanical systems with Coulomb friction, stiction, impact and constraint addition deletion—I theory. Mechanism and Machine Theory 21(5):401–406.

Hunt KH, Crossley FR (1975) Coefficient of restitution interpreted as damping in vibroimpact. Journal of Applied Mechanics 7:440–445.

Hurmuzlu Y, Marghitu DB (1994) Rigid body collision of planar kinematic chain with multiple contact points. The International Journal of Robotics Research 13:82–89.

Huston RL (1990) Multibody dynamics. Butterworth-Heinemann, Boston, MA.

Huston RL (1991) Multibody dynamics—modeling and analysis methods. Applied Mechanics Review 44(3):109–117.

Innocenti C (2002) Kinematic clearance sensitivity analysis of spatial structures with revolute joints. Journal of Mechanical Design 124:52–57.

Jiménez JM, Avello A, García-Alonso A, Garcia de Jálon J (1990) COMPAMM—a simple and efficient code for kinematic and dynamic numerical simulation of 3-D multibody system with realistic graphics. Multibody systems handbook. Springer, Berlin Heidelberg New York.

Johnson KL (1999) Contact mechanics. Cambridge University Press, Cambridge.

Kakizaki T, Deck JF, Dubowsky S (1993) Modeling the spatial dynamics of robotic manipulators with flexible links and joint clearances. Journal of Mechanical Design 115:839–847.

Kane TR (1984) A dynamic puzzle. Stanford Mechanics Alumni Club Newsletter, p. 6.

Kane TR, Levinson DA (1985) Dynamics: theory and applications. McGraw-Hill, New York.

Kaufman RE (1978) Mechanism design by computer. Machine Design 50(24):94–100.

Keller JB (1986) Impact with friction. Journal of Applied Mechanics 53:1–4.

Khulief YA, Shabana AA (1986) Dynamic analysis of constrained system of rigid and flexible bodies with intermittent motion. Journal of Mechanisms, Transmissions, and Automation in Design 108:38–45.

Khulief YA, Shabana AA (1987) A continuous force model for the impact analysis of flexible multibody systems. Mechanism and Machine Theory 22(3):213–224.

Kilmister N, Reeve JE (1966) Rotational mechanics. Longmans, London.

Kreuzer E (1979) Symbolische Berechnung der Bewegungsgleichungen von Mehrkörpersystemem. Number 32 Fortschritt-Berichte VDI, Reihe 11. VDI Verlag, Düsseldorf, Deutschland.

Lankarani HM (1988) Canonical equations of motion and estimation of parameters in the analysis of impact problems. Ph.D. Dissertation, University of Arizona, Tucson, AZ.

Lankarani HM, Nikravesh PE (1990) A contact force model with hysteresis damping for impact analysis of multibody systems. Journal of Mechanical Design 112:369–376.

Lee TW, Wang AC (1983) On the dynamics of intermittent-motion mechanisms. Part 1—dynamic model and response. Journal of Mechanisms, Transmissions, and Automation in Design 105:534–540.

Liu TS, Lin YS (1990) Analysis of flexible linkages with lubricated joints. Journal of Sound and Vibration 141(2):193–205.

Lötstedt P (1982) Mechanical systems of rigid bodies subjected to unilateral constraints. Computer Graphics 23:223–232.

Mahrus D (1974) Experimental investigation into journal bearing performance. Revista Brasileira de Tecnologia 5(3–4):139–152.

Martin GH (1982) Kinematics and dynamics of machines. McGraw-Hill, New York.

Maw N, Barber JR, Fawcett JN (1976) The oblique impact of elastic spheres. Wear 101–114.

Miedema B, Mansour WM (1976) Mechanical joints with clearance: a three mode model. Journal of Engineering for Industry 98(4):1319–1323.

Moes H, Sikkes EG, Bosma R (1986) Mobility and impedance tensor methods for full and partial-arc journal bearings. Journal of Tribology 108:612–620.

Moreau JJ (1988) Unilateral contact and dry friction in finite freedom dynamics. Non-smooth mechanics and applications, CISM courses and lectures, edited by JJ Moreau and PP Panagiotopoulos, Vol. 302, pp. 1–82. Springer, Berlin Heidelberg New York.

Neto MA, Ambrósio J (2003) Stabilization methods for the integration of DAE in the presence of redundant constraints. Multibody System Dynamics 10:81–105.

Newton I (1686) Philosophia naturalis principia mathematica. S. Pepsy, Reg. Soc, London.

Nikravesh PE (1988) Computer-aided analysis of mechanical systems. Prentice Hall, Englewood Cliffs, NJ.

Norton RL, Ault HK, Wiley J, Parks T, Calawa R, Wickstrand M (1994) Bearing forces as function of mechanical stiffness and vibration isolation in a four bar linkage. ASTM Special Technical Publication Proceedings of the symposium on effects of mechanical stiffness and vibration on wear Nr 1247, Montreal, Canada.

Orlandea N, Chace MA, Calahan DA (1977) A sparsity oriented approach to the dynamic analysis and design of mechanical systems—parts 1 and 2. Journal of Engineering for Industry 99:773–784.

Painlevé P (1905) Sur les lois de frottement de glissement. Comptes Rendus de l'Academie des Sciences Paris 121:112–115; 141:401–405; 141:546–552.

Pang JS, Cottle RW, Stone RE (1992) The linear complementary problem. Academic Press, Boston, MA.

Paul B, Krajcinovic D (1970) Computer analysis of machines with planar motion, part 1—kinematics, part 2—dynamics. Journal of Applied Mechanics 37:697–712.

Pfeiffer F (1999) Unilateral problems of dynamics. Archive of Applied Mechanics 69:503–527.

Pfeifer F, Glocker C (1996) Multibody dynamics with unilateral constraints. Wiley, New York.

Pinkus O, Sternlicht SA (1961) Theory of hydrodynamic lubrication. McGraw-Hill, New York.

Pombo JCEJ (2004) A multibody methodology for railway dynamics applications. Ph.D. Dissertation, Universidade Técnica de Lisboa, Instituto Superior Técnico, Lisbon, Portugal.

Rahnejat H (2000) Multi-body dynamics: historical evolution and application. Proceedings of the Institution of Mechanical Engineers, Journal of Mechanical Engineering Science 214(C):149–173.

Ravn P (1998) A continuous analysis method for planar multibody systems with joint clearance. Multibody System Dynamics 2:1–24.

RecurDyn (2006) RecurDyn version 6.3, Release notes. Functionbay, Inc.

Reuleaux F (1963) The kinematics of machinery. Dover, New York.

Rhee J, Akay A (1996) Dynamic response of a revolute joint with clearance. Mechanism and Machine Theory 31(1):121–124.

Rockafellar RT (1972) Convex analysis. Princeton University Press, New Jersey.

Roger RJ, Andrews GC (1977) Dynamic simulation of planar mechanical systems with lubricated bearing clearances using vector-network methods. Journal of Engineering for Industry 99(1):131–137.

Rosenberg RM (1977) Analytical dynamics of discrete systems. Plenum Press, New York.

Routh ET (1905) Dynamics of a system of rigid bodies. Macmillan, London.

Rubel AJ, Kaufman RE (1977) KINSYN III: a new human-engineering system for interactive computer-aided design of planar linkages. Journal of Engineering for Industry, Series B 99(2):440–448.

Rulka W (1990) SIMPACK—a computer program for simulation of large motion multibody systems. Multibody systems handbook, edited by W Schichlen, pp. 265–284. Springer, Berlin Heidelberg New York.

Ryan RR (1990) ADAMS—multibody system analysis software. Multibody systems handbook. Springer, Berlin Heidelberg New York.

Schiehlen W (1990) Multibody systems handbook. Springer, Berlin Heidelberg New York.

Schiehlen W (1997) Multibody system dynamics: roots and perspectives. Multibody System Dynamics 1:149–188.

Schwab AL (2002) Dynamics of flexible multibody systems, small vibrations superimposed on a general rigid body motion. Ph.D. Dissertation, Delft University of Technology, The Netherlands.

Seneviratne LD, Earles SWE (1992) Chaotic behaviour exhibited during contact loss in a clearance joint in a four-bar mechanism. Mechanism and Machine Theory 27(3):307–321.

Shabana AA (1997) Flexible multibody dynamics: review of past and recent developments. Multibody System Dynamics 1:189–222.

Shabana AA (1989) Dynamics of multibody systems. Wiley, New York.

Shampine L, Gordon M (1975) Computer solution of ordinary differential equations: the initial value problem. Freeman, San Francisco, CA.

Sheth PN, Uicker JJ (1971) IMP (Integrated Mechanism Program): a computer-aided design analysis system for mechanisms and linkages. Journal of Engineering for Industry, Series B 94(2):454–464.

Shigley JE, Uicker JJ (1995) Theory of machines and mechanisms. McGraw-Hill, New York.

Smith CE (1991) Predicting rebounds using rigid-body dynamics. Journal of Applied Mechanics 58:754–758.

Soong K (1988) An analytical and experimental study of the elastodynamic response characteristics of planar linkage mechanism with bearing clearances. Ph.D. Dissertation, Michigan State University, Flint.

Soong K, Thompson BS (1990) A theoretical and experimental investigation of the dynamic response of a slider–crank mechanism with radial clearance in the gudgeon–pin joint. Journal of Mechanical Design 112:183–189.

Stronge WJ (1991) Unraveling paradoxical theories for rigid body collisions. Journal of Applied Mechanics 58:1049–1055.

Stronge WJ (2000) Impact mechanics. Cambridge University Press, Cambridge.

Timoshenko SP, Goodier JN (1970) Theory of elasticity. McGraw-Hill, New York.

TNO (1998) MADYMO Theoretical manual, Version 5.3. TNO, Delft, The Netherlands.

Townsend MA, Mansour WM (1975) A pendulating model for mechanisms with clearances in the revolutes. Journal of Engineering for Industry, Series B 97(2):354–358.

Trinkle JC (2003) Formulation of multibody dynamics as complementary problems. Proceedings of 2003 ASME design engineering technical conferences, Chicago, IL, pp. 361–370.

Uicker JJ (1969) Dynamic behavior of spatial linkages, part 1—exact equations of motion, part 2—small oscillations about equilibrium. Journal of Engineering for Industry 91:251–265.

Wang Y, Mason MT (1992) Two dimensional rigid-body collisions with friction. Journal of Applied Mechanics 59:635–642.

Whittaker ET (1904) A treatise on the analytical dynamics of particles and rigid bodies. Cambridge University Press, Cambridge.

Wilson R, Fawcett JN (1974) Dynamics of slider–crank mechanism with clearance in the sliding bearing. Mechanism and Machine Theory 9:61–80.

Winfrey RC, Anderson RV, Gnilka CW (1973) Analysis of elastic machinery with clearances. Journal of Engineers for Industry 95:695–703.

Wittenburg J (1977) Dynamics of systems of rigid bodies. B.G. Teubner, Stuttgart, Germany.

Wu CLS, Earles SWE (1977) A determination of contact-loss at a bearing of a linkage mechanism. Journal of Engineering for Industry, Series B 99(2): 375–380.

Zukas JA, Nicholas T, Greszczuk LB, Curran DR (1982) Impact dynamics. Wiley, New York.

Chapter 2
Multibody Systems Formulation

The dynamic analysis of multibody systems, made of interconnected bodies that undergo large displacements and rotations, is a research area with applications in a broad variety of engineering fields that deserved relevant attention over the last few decades (Wittenburg 1977, Nikravesh 1988, Huston 1990, Schiehlen 1990). Many multibody computational programs capable of automatic generation and integration of the differential equations of motion have been developed, such as DAP (Nikravesh 1988), DADS (Smith and Haug 1990), ADAMS (Ryan 1990), COM-PAMM (Jiménez et al. 1990) and SIMPACK (Rulka 1990). The various formulations of multibody systems used in these programs differ in the principle used (e.g., principle of virtual work, principle of virtual power, Newton–Euler's approach), types of coordinates adopted (e.g., Cartesian coordinates, Lagrangian coordinates) and the methods selected for handling kinematic constraints (e.g., coordinate partitioning method, augmented Lagrange formulation) (Nikravesh 1988, Shabana 1989). The solution of the problem for constrained multibody systems can be obtained using the Lagrange's multipliers technique which leads to a set of differential and algebraic equations (DAE), in which the coordinates and the Lagrange multipliers are unknown. The numerical solution of the set of DAE is not straightforward (Brenan et al. 1989). One of the most used methods to solve this problem consists of converting the system of DAE into a set of ordinary differential equations (ODE) by appending the second derivative with respect to time of the constraint equations (Gear 1981, Nikravesh 1988). However, this approach presents numerical stability problems, which can be reduced or at least kept under control by employing a technique of constraint violation stabilization (Baumgarte 1972). The formulation of multibody system dynamics adopted in this work follows closely that of Nikravesh (1988), in which the generalized Cartesian coordinates and the Newton–Euler's approach are employed to obtain the equations of motion. Furthermore, the Baumgarte stabilization technique is used to control the position and velocity violations (Baumgarte 1972). The methodology presented is implemented in the computational program DAP, which has been developed for dynamic analysis of general multibody systems (Nikravesh 1988). This multibody formulation is capable of automatically generating and solving the equations of motion for general multibody systems.

P. Flores et al., *Kinematics and Dynamics of Multibody Systems with Imperfect Joints.*
© Springer 2008

2.1 Multibody System Definition

In order to analyze the dynamic response of a constrained multibody system (MBS), it is first necessary to formulate the equations of motion that govern its behavior. Figure 2.1 depicts a multibody system, which consists of a collection of rigid and/or flexible bodies interconnected by kinematic joints and possibly some force elements. The kinematic joints control the relative motion between the bodies, while the force elements represent the internal forces that develop between bodies due to their relative motion. The forces applied over the system components may be the result of springs, dampers, actuators or external forces. A wide variety of mechanical systems, such as robots, heavy machinery, automobile or railway vehicles, can be modeled in this way (Nikravesh 1988). In the plane, the rigid bodies have three degrees of freedom (DOF). When the bodies include some flexibility they have three DOF plus the number of generalized flexible coordinates necessary to describe their deformation (Ambrósio 1991). Here, only rigid bodies are used.

Kinematic constraints can be classified as holonomic or nonholonomic. Holonomic constraints arise from geometric constraints and are integrable into a form involving only coordinates (*holo* comes from Greek that means whole, integer). Nonholonomic constraints are not integrable. The relationship specified by a constraint can be an explicit function of time designated as rheonomic constraints (*rheo* comes from Greek that means hard, inflexible, independent) or not, being designated by scleronomic constraints (*scleros* comes from Greek that means flexible, changing).

The kinematic constraints considered here are assumed to be holonomic, arising from geometric constraints on the generalized coordinates. Holonomic constraints,

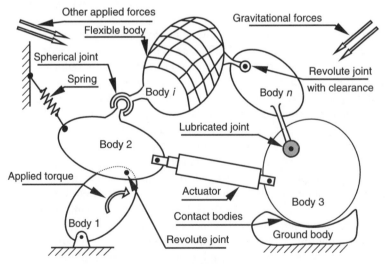

Fig. 2.1 Schematic representation of a general multibody system made of rigid bodies interconnected by kinematic joints and acted on by applied forces

also called geometric restrictions, are algebraic equations imposed on the system that are expressed as functions of the displacement and, possibly, time. If the time t does not appear explicitly in the constraint equation, then the system is said to be scleronomic. A simple example of scleronomic constraint equations is the revolute joint between two bodies. Otherwise when the constraint is holonomic and t appears explicitly, the system is said to be rheonomic. When the constraint equation contains inequalities or relations between velocity components that are not integrable in closed form, the system is said to be nonholonomic. Nonholonomic constraints are still kinematic constraints since they impose restrictions to the velocity and acceleration (Nikravesh 1988, Shabana 1989).

With respect to the selection of coordinates two main approaches are generally found in the methods used in the dynamic analysis of MBS. Either an expanded system of dependent coordinates, e.g., Cartesian coordinates, are used to describe the system configuration or a minimum number of relative coordinates are used, corresponding to the mechanical degrees of freedom of the system and lead to a minimum number of ordinary differential equations (Shabana 1989). In relative coordinates, the kinematic constraints are implicit for open-loop systems. In the presence of closed-loop systems, the use of relative coordinates requires that either a joint is described by a pair of forces, i.e., a cut-joint, or that kinematic constraints are set explicitly in one of the joints of the closed loop. The application of Cartesian coordinates has the advantage that the formulation of the equations of motion even for complex systems is straightforward (Nikravesh 1988).

2.2 Cartesian Coordinates

With the purpose of defining the geometric configuration or position of a multibody system, it is necessary to specify the coordinates that locate each body. Due to their simplicity and computational easiness, Cartesian coordinates are used to formulate the dynamics of the multibody systems. Figure 2.2 represents a rigid body i to which a body-fixed coordinate system $(\xi\eta)_i$ is attached at its center of mass. When Cartesian coordinates are used, the position and orientation of the rigid body is defined by a set of translational and rotational coordinates. Thus body i is uniquely located in the plane by specifying the global coordinates $\mathbf{r}_i = [x \quad y]_i^T$ of the body-fixed coordinate system origin and the angle θ_i of rotation of this system relative to the global coordinate system.

A point P on body i can be defined by position vector $\mathbf{s}_i^P$, which represents the location of point P with respect to the body-fixed reference frame $(\xi\eta)_i$, and by the global position vector $\mathbf{r}_i$, that is,

$$\mathbf{r}_i^P = \mathbf{r}_i + \mathbf{s}_i^P = \mathbf{r}_i + \mathbf{A}_i \mathbf{s}_i'^P \tag{2.1}$$

where $\mathbf{A}_i$ is the transformation matrix for body i that defines the orientation of the referential $(\xi\eta)_i$ with respect to the referential frame XY.

Fig. 2.2 Rigid body in
Cartesian coordinates

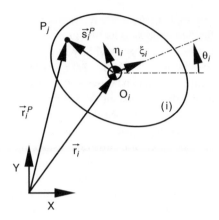

The transformation matrix is expressed as

$$\mathbf{A}_i = \begin{bmatrix} \cos\theta & -\sin\theta \\ \sin\theta & \cos\theta \end{bmatrix}_i \tag{2.2}$$

Thus the vector of coordinates for body i is denoted by

$$\mathbf{q}_i = [\mathbf{r}^T \ \theta]_i^T \tag{2.3}$$

Note that alternative sets of coordinates can be in the literature. The relative advantages and drawbacks are often application dependent and, consequently, the problem of choosing the most efficient type of coordinates for each type of application is not discussed any further in this text.

2.3 Constraints for Revolute and Translational Joints

A kinematic joint imposes limitations on the relative motion between adjacent bodies. When these conditions are expressed in analytical form, they are called constraint equations. In a simple way, a constraint is any condition that reduces the number of degrees of freedom in a system. In what follows, the formulation for the planar revolute and translational joints is reviewed to illustrate the methodology. With these two basic kinematic joints a large class of multibody systems can be modeled and analyzed. For details on the formulation of other types of kinematic joints the interested reader is referred to Nikravesh (1988). In order to distinguish among the different constraint equations, each elementary set of constraints is identified by a superscript containing two parameters. The first parameter denotes the type of constraint while the second one defines the number of independent equations that it involves. For example, $\mathbf{\Phi}^{(r,2)}$ denotes the planar revolute (r) joint constraint, which contains two (2) equations.

The revolute joint is a pin and bush type of joint that constrains the relative translation between the two bodies i and j, allowing only the relative rotations, as illustrated in Fig. 2.3. The kinematic conditions for the revolute joint require that two different points, each one belonging to a different body, share the same position in space all the time. This means that the global position of a point P in body i is coincident with the global position of a point P in body j. Such a condition is expressed by two algebraic equations that can be obtained from the following vector loop equation:

$$\mathbf{r}_i + \mathbf{s}_i^P - \mathbf{r}_j - \mathbf{s}_j^P = \mathbf{0} \tag{2.4}$$

which is re-written as

$$\mathbf{\Phi}^{(r,2)} \equiv \mathbf{r}_i + \mathbf{A}_i \mathbf{s}_i'^P - \mathbf{r}_j - \mathbf{A}_j \mathbf{s}_j'^P = \mathbf{0} \tag{2.5}$$

Thus there is only one relative DOF between two bodies that are connected by a planar revolute joint.

In a planar translational joint the two bodies, slider and guide, translate with respect to each other parallel to the line of translation, so that there is neither rotation between the bodies nor a relative translation motion in the direction perpendicular to the line of translation. Thus a planar translational joint reduces the number of degrees of freedom of the system by two, which implies the need for two independent algebraic equations to represent it.

A constraint equation for eliminating the relative rotation between two bodies i and j is written as

$$\theta_i - \theta_j - (\theta_i^0 - \theta_j^0) = 0 \tag{2.6}$$

where θ_i^0 and θ_j^0 are the initial rotational angles for each body. In order to eliminate the relative motion between the two bodies in a direction perpendicular to the line of translation, the two vectors $\mathbf{s}_i$ and $\mathbf{d}$, shown in Fig. 2.4, must remain parallel. These vectors are defined by locating three points on the line of translation, two

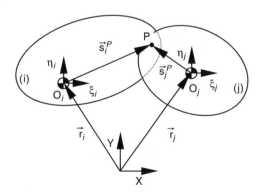

Fig. 2.3 Planar revolute joint connecting bodies i and j

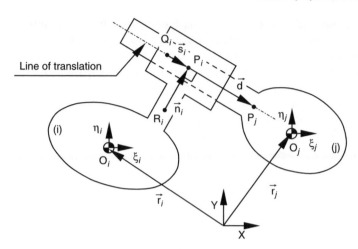

Fig. 2.4 Planar translational joint connecting bodies i and j

points on body i and one point on body j. This condition is enforced by forcing the vector product of these two vectors to remain null all the time. A simple method consists in defining another vector $\mathbf{n}_i$ fixed to body i and perpendicular to the line of translation. Then it is only required that $\mathbf{d}$ remains perpendicular to $\mathbf{n}_i$, that is,

$$\mathbf{n}_i^T \mathbf{d} = 0 \tag{2.7}$$

Therefore (2.6) and (2.7) yield the two constraint equations for a planar translational joint:

$$\mathbf{\Phi}^{(t,2)} = \begin{bmatrix} (x_i^P - x_i^Q)(y_j^P - y_i^P) - (y_i^P - y_i^Q)(x_j^P - x_i^P) \\ \theta_i - \theta_j - (\theta_i^0 - \theta_j^0) \end{bmatrix} = \begin{bmatrix} 0 \\ 0 \end{bmatrix} \tag{2.8}$$

where x_i^P and y_i^P are the global coordinates of point P of body i, and the coordinates of points Q on body i and P on body j follow the same notation.

2.4 Kinematic Analysis

Kinematic analysis is the study of the motion of a system independently of the forces that produce it. Since in the kinematic analysis the forces are not considered, the motion of the system is specified by driving elements that govern the motion of specific degrees of freedom of the system during the analysis. The position, velocity and acceleration of the remaining elements of the system are defined by kinematic constraint equations that describe the system topology. It is clear that in the kinematic analysis, the number of driver constraints must be equal to the number of degrees of freedom of the multibody mechanical system. In short, the kinematic analysis is performed by solving a set of equations that result from the kinematic and driver constraints.

Let the configuration of a MBS be described by nc Cartesian coordinates. A set of m algebraic kinematic independent holonomic constraints $\mathbf{\Phi}$ can be written in a compact form as (Nikravesh 1988)

$$\mathbf{\Phi}(\mathbf{q}, t) = \mathbf{0} \tag{2.9}$$

where $\mathbf{q}$ is the vector of generalized coordinates and t is the time variable, in general associated with the driving elements.

The velocities and accelerations of the system elements are evaluated using the velocity and acceleration constraint equations. Thus the first time derivative of (2.9) provides the velocity constraint equations

$$\mathbf{\Phi_q \dot{q}} = -\mathbf{\Phi}_t \equiv \upsilon \tag{2.10}$$

where $\mathbf{\Phi_q}$ is the Jacobian matrix of the constraint equations, that is, the matrix of the partial derivates, $\partial\mathbf{\Phi}/\partial\mathbf{q}$, $\dot{\mathbf{q}}$ is the vector of generalized velocities and υ is the right-hand side of velocity equations, which contain the partial derivates of $\mathbf{\Phi}$ with respect to time, $\partial\mathbf{\Phi}/\partial t$. Notice that only rheonomic constraints, associated with driver equations, contribute with nonzero entries to the vector υ.

A second differentiation of (2.9) with respect to time leads to the acceleration constraint equations, obtained as

$$\mathbf{\Phi_q \ddot{q}} = -(\mathbf{\Phi_q \dot{q}})_{\mathbf{q}}\dot{\mathbf{q}} - 2\mathbf{\Phi_{qt}}\dot{\mathbf{q}} - \mathbf{\Phi}_{tt} \equiv \boldsymbol{\gamma} \tag{2.11}$$

where $\ddot{\mathbf{q}}$ is the acceleration vector and γ is the right-hand side of acceleration equations, i.e., the vector of quadratic velocity terms, which are exclusively function of velocity, position and time. In the case of scleronomic constraints, that is, when $\mathbf{\Phi}$ is not explicitly dependent on time, the term $\mathbf{\Phi}_t$ in (2.9) and $\mathbf{\Phi_{qt}}$ and $\mathbf{\Phi}_{tt}$ in (2.11) vanish.

The constraint equations represented by (2.9) are nonlinear in terms of $\mathbf{q}$ and are, usually, solved by employing the Newton–Raphson method. Equations (2.10) and (2.11) are linear in terms of $\dot{\mathbf{q}}$ and $\ddot{\mathbf{q}}$, respectively, and can be solved by any usual method adopted for the solution of systems of linear equations.

The kinematic analysis of a multibody system can be carried out by solving (2.9), (2.10) and (2.11). The necessary steps to perform this analysis, sketched in Fig. 2.5, are summarized as follows:

1. Specify the initial conditions for positions $\mathbf{q}^0$ and initialize the time counter t^0.
2. Evaluate the position constraint (2.9) and solve them for positions, $\mathbf{q}$.
3. Evaluate the velocity constraint (2.10) and solve them for velocities, $\dot{\mathbf{q}}$.

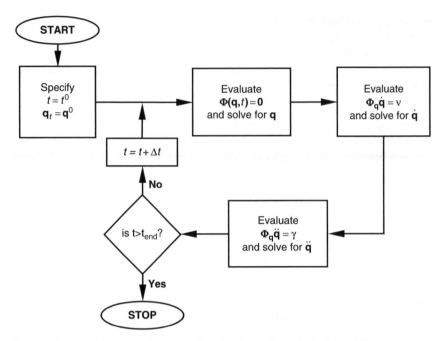

Fig. 2.5 Flowchart of computational procedure for kinematic analysis of a multibody system

4. Evaluate the acceleration constraint (2.11) and solve them for accelerations, $\ddot{\mathbf{q}}$.
5. Increment the time. If the time is smaller than the final time, go to step 2), otherwise stop the kinematic analysis.

2.5 Equations of Motions for a Constrained Multibody System

The equations of motion for a constrained multibody system of rigid bodies are written as (Nikravesh 1988)

$$\mathbf{M}\ddot{\mathbf{q}} = \mathbf{g} + \mathbf{g}^{(c)} \qquad (2.12)$$

where $\mathbf{M}$ is the system mass matrix, $\ddot{\mathbf{q}}$ is the vector that contains the system accelerations, $\mathbf{g}$ is the generalized force vector, which contains all external forces and moments, and $\mathbf{g}^{(c)}$ is the vector of constraint reaction equations.

The joint reaction forces can be expressed in terms of the Jacobian matrix of the constraint equations and the vector of Lagrange multipliers as (Nikravesh 1988)

$$\mathbf{g}^{(c)} = -\boldsymbol{\Phi}_{\mathbf{q}}^{T}\boldsymbol{\lambda} \qquad (2.13)$$

where $\boldsymbol{\lambda}$ is the vector that contains m unknown Lagrange multipliers associated with m holonomic constraints. The Lagrange multipliers are physically related to the reaction forces and moments generated between the bodies interconnected by kinematic joints. Thus substitution of (2.13) in (2.12) yields

$$\mathbf{M}\ddot{\mathbf{q}} + \mathbf{\Phi}_{\mathbf{q}}^{T}\boldsymbol{\lambda} = \mathbf{g} \tag{2.14}$$

In dynamic analysis, a unique solution is obtained when the constraint equations are considered simultaneously with the differential equations of motion, for a proper set of initial conditions. Therefore (2.11) is appended to (2.14), yielding a system of differential algebraic equations (DAE). Mathematically the simulation of constrained multibody system requires the solution of a set of nc differential equations coupled with a set of m algebraic equation, written as

$$\begin{bmatrix} \mathbf{M} & \mathbf{\Phi}_{\mathbf{q}}^{T} \\ \mathbf{\Phi}_{\mathbf{q}} & \mathbf{0} \end{bmatrix} \begin{Bmatrix} \ddot{\mathbf{q}} \\ \boldsymbol{\lambda} \end{Bmatrix} = \begin{Bmatrix} \mathbf{g} \\ \boldsymbol{\gamma} \end{Bmatrix} \tag{2.15}$$

This system of equations is solved for $\ddot{\mathbf{q}}$ and $\boldsymbol{\lambda}$. Then, in each integration time step, the accelerations vector, $\ddot{\mathbf{q}}$, together with velocities vector, $\dot{\mathbf{q}}$, are integrated in order to obtain the system velocities and positions for the next time step. This procedure is repeated till the final analysis time is reached.

The system of (2.15) can be solved by applying any method suitable for the solution of linear algebraic equations. The existence of null elements in the main diagonal of the matrix and the possibility of ill-conditioned matrices suggest that methods using partial or full pivoting are preferred. However, none of these formulations help in the presence of redundant constraints.

The left-hand-side matrix of the system of (2.15) can be inverted analytically. For the purpose, (2.14) is rearranged to put the acceleration vector in evidence in the left-hand side and the result is substituted in (2.11), which is also rearranged as

$$\boldsymbol{\lambda} = (\mathbf{\Phi}_{\mathbf{q}}\mathbf{M}^{-1}\mathbf{\Phi}_{\mathbf{q}}^{T})^{-1}\mathbf{\Phi}_{\mathbf{q}}\mathbf{M}^{-1}\mathbf{g} - (\mathbf{\Phi}_{\mathbf{q}}\mathbf{M}^{-1}\mathbf{\Phi}_{\mathbf{q}}^{T})^{-1}\boldsymbol{\gamma} \tag{2.16}$$

In these equations, it is assumed that the multibody model does not include any body with null mass or inertia so that the inverse of the mass matrix $\mathbf{M}$ exists. The substitution of (2.16) in (2.14) provides the expression for the system accelerations written as

$$\ddot{\mathbf{q}}^{*} = \left[\mathbf{M}^{-1} - \mathbf{M}^{-1}\mathbf{\Phi}_{\mathbf{q}}^{T}(\mathbf{\Phi}_{\mathbf{q}}\mathbf{M}^{-1}\mathbf{\Phi}_{\mathbf{q}}^{T})^{-1}\mathbf{\Phi}_{\mathbf{q}}\mathbf{M}^{-1}\right]\mathbf{g} + \mathbf{M}^{-1}\mathbf{\Phi}_{\mathbf{q}}^{T}(\mathbf{\Phi}_{\mathbf{q}}\mathbf{M}^{-1}\mathbf{\Phi}_{\mathbf{q}}^{T})^{-1}\boldsymbol{\gamma} \tag{2.17}$$

Equations (2.16) and (2.17) are now rearranged in a compact form as

$$\begin{bmatrix} \ddot{\mathbf{q}}^{*} \\ \boldsymbol{\lambda} \end{bmatrix} = \begin{bmatrix} \left[\mathbf{M}^{-1} - \mathbf{M}^{-1}\mathbf{\Phi}_{\mathbf{q}}^{T}(\mathbf{\Phi}_{\mathbf{q}}\mathbf{M}^{-1}\mathbf{\Phi}_{\mathbf{q}}^{T})^{-1}\mathbf{\Phi}_{\mathbf{q}}\mathbf{M}^{-1}\right] & \mathbf{M}^{-1}\mathbf{\Phi}_{\mathbf{q}}^{T}(\mathbf{\Phi}_{\mathbf{q}}\mathbf{M}^{-1}\mathbf{\Phi}_{\mathbf{q}}^{T})^{-1} \\ (\mathbf{\Phi}_{\mathbf{q}}\mathbf{M}^{-1}\mathbf{\Phi}_{\mathbf{q}}^{T})^{-1}\mathbf{\Phi}_{\mathbf{q}}\mathbf{M}^{-1} & -(\mathbf{\Phi}_{\mathbf{q}}\mathbf{M}^{-1}\mathbf{\Phi}_{\mathbf{q}}^{T})^{-1} \end{bmatrix} \begin{bmatrix} \mathbf{g} \\ \boldsymbol{\gamma} \end{bmatrix} \tag{2.18}$$

The matrix in the right-hand side of (2.18) is the inverse of the system matrix that appears in (2.15).

The system of the motion (2.15) does not explicitly use the position and velocity equations associated with the kinematic constraints, that is, (2.9) and (2.10). Consequently, for moderate or long simulations, the original constraint equations start to be violated due to the integration process and/or the inaccurate initial conditions. Therefore special procedures must be followed to avoid or minimize this phenomenon. Several methods to solve this problem have been suggested and tested, the most common among them being the Baumgarte stabilization method (Baumgarte 1972), the coordinate partitioning method (Wehage and Haug 1982) and the augmented Lagrangian formulation (Bayo et al. 1988).

In addition to these three basic approaches, many research papers have been published on the stabilization methods for the numerical integration of the equations of motion of multibody systems, namely, Park and Chiou (1988), Kim et al. (1990), Yoon et al. (1994), Rosen and Edelstein (1997), Lin and Hong (1998), Blajer (1999), Lin and Huang (2002). Nikravesh (1984) made a comparative study of the standard or direct integration of the system's equation of motion, the Baumgarte's approach and the coordinate partitioning method. More recently, Neto and Ambrósio (2003) used different methodologies to handle the constraint violation correction or stabilization for the integration of DAE in the presence of redundant constraints discussing, in the process, the benefits and shortcoming of these methods.

Due to its simplicity and easiness for computational implementation, the Baumgarte stabilization method (BSM) is the most popular and attractive technique to control constraint violations. However, this method does not solve all possible numerical difficulties as, for instance, those that arise near kinematic singularities. Another drawback of Baumgarte's method is the ambiguity in choosing feedback parameters. The choice of these coefficients usually involves a trial-and-error procedure (Baumgarte 1972). The augmented Lagrangian method, which also keeps the constraint violations under control, shares with BSM the problem of parameter selection but is able to handle redundant constraints. These methods are discussed in detail in Sect. 2.7.

2.6 Direct Integration Method of the Equations of Motion

In this section, the main numerical aspects related to the standard integration of the equations of motion of a multibody system are reviewed. The standard integration of the equations of motion, here called direct integration method (DIM), converts the nc second-order differential equations of motion into $2nc$ first-order differential equations. Then a numerical scheme, such as the Runge–Kutta method, is employed to solve the initial-value problem (Gear 1981).

The $2nc$ differential equations of motion are solved, without considering the integration numerical errors and, consequently, during the simulation the propagation of these kinds of errors results in constraint violations. The two error sources that lead to constraint violations for any numerical integration step are truncation and round-off errors. Truncation or discretization errors are caused by the nature of the techniques employed to approximate values of a function $\mathbf{y}$. Round-off errors are due to the limited numbers of significant digits that can be retained by a computer. Truncation errors are composed of two parts. The first one is a local truncation error that results from an application of the method in question over a single step. The second one is a propagated error that results from the approximation produced during the previous steps. The sum of the two is the total or global truncation errors. The commonly used numerical integration algorithms are useful in solving first-order differential equations that take the form (Gear 1981)

$$\dot{\mathbf{y}} = f(\mathbf{y}, t) \tag{2.19}$$

Thus, if there are nc second-order differential equations, they are converted to $2nc$ first-order equations by defining the $\mathbf{y}$ and $\dot{\mathbf{y}}$ vectors, which contain, respectively, the system positions and velocities and the system velocities and accelerations, as follows:

$$\mathbf{y} = \left\{ \begin{array}{c} \mathbf{q} \\ \dot{\mathbf{q}} \end{array} \right\} \quad \text{and} \quad \dot{\mathbf{y}} = \left\{ \begin{array}{c} \dot{\mathbf{q}} \\ \ddot{\mathbf{q}} \end{array} \right\} \tag{2.20}$$

The reason for introducing the new vectors $\mathbf{y}$ and $\dot{\mathbf{y}}$ is that most numerical integration algorithms deal with first-order differential equations (Shampine and Gordon 1975). The following diagram can interpret the process of numerical integration at instant of time t:

$$\dot{\mathbf{y}}(t) \xrightarrow{\textit{Integration}} \mathbf{y}(t + \Delta t) \tag{2.21}$$

In other words, velocities and accelerations at instant t, after integration process, yield positions and velocities at next time step, $t = t + \Delta t$.

Figure 2.6 presents a flowchart that shows the algorithm of direct integration method of the equations of motion. At $t=t^0$, the initial conditions on $\mathbf{q}^0$ and $\dot{\mathbf{q}}^0$ are required to start the integration process. These values cannot be specified arbitrarily, but must satisfy the constraint equations defined by (2.9) and (2.10). The direct integration algorithm presented in Fig. 2.6 can be summarized by the following steps:

1. Start at instant of time t^0 with given initial conditions for positions $\mathbf{q}^0$ and velocities $\dot{\mathbf{q}}^0$.
2. Assemble the global mass matrix $\mathbf{M}$, evaluate the Jacobian matrix $\mathbf{\Phi_q}$, construct the constraint equations $\mathbf{\Phi}$, determine the right-hand side of the accelerations $\mathbf{\gamma}$ and calculate the force vector $\mathbf{g}$.
3. Solve the linear set of the equations of motion (2.15) for a constrained multibody system in order to obtain the accelerations $\ddot{\mathbf{q}}$ at time t and the Lagrange multipliers $\mathbf{\lambda}$.

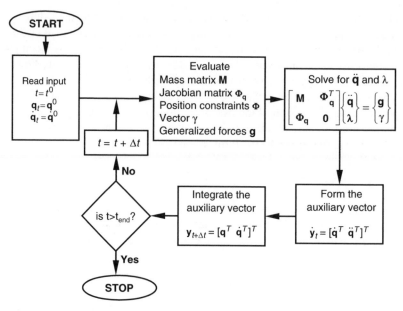

Fig. 2.6 Flowchart of computational procedure for dynamic analysis of multibody systems; direct integration method (DIM)

4. Assemble the vector $\dot{\mathbf{y}}_t$ containing the generalized velocities $\dot{\mathbf{q}}$ and accelerations $\ddot{\mathbf{q}}$ for instant of time t.
5. Integrate numerically the $\dot{\mathbf{q}}$ and $\ddot{\mathbf{q}}$ for time step $t + \Delta t$ and obtain the new positions and velocities.
6. Update the time variable, go to step 2) and proceed with the process for a new time step, until the final time of analysis is reached.

The direct integration method of equations of motion is prone to integration errors, because the constraint (2.9) and (2.10) are only satisfied at the initial instant of time. In the first few time steps, the constraint violations are usually small and negligible. However, as time progresses, the error in computed values for kinematic parameters is accumulated and constraint violations increase. Hence the results produced can be unacceptable; therefore the DIM requires the use of a constraint stabilization technique, especially for long simulations. It should be noted that the DIM is quite sensitive to initial conditions, which can be an important source of errors in the integration process.

2.7 Solution Methods that Handle Constraint Violations

The initial conditions and the integration of the velocities and accelerations of the multibody system introduce numerical errors in the new positions and velocities obtained. These errors are due to the finite precision of the numerical methodologies and the position and velocity constraint equations not appearing anywhere in the solution procedure outlined in Fig. 2.6. Therefore methods able to eliminate errors in the constraint or velocity equations or, at least, to keep such errors under control must be implemented (Neto and Ambrósio 2003).

2.7.1 Baumgarte Stabilization Method

As shown earlier, for a multibody system of kinematically constrained rigid bodies, the equations of motion are given by

$$\begin{bmatrix} \mathbf{M} & \mathbf{\Phi}_q^T \\ \mathbf{\Phi}_q & \mathbf{0} \end{bmatrix} \begin{Bmatrix} \ddot{\mathbf{q}} \\ \lambda \end{Bmatrix} = \begin{Bmatrix} \mathbf{g} \\ \gamma \end{Bmatrix} \tag{2.22}$$

with initial conditions for $\mathbf{q}^0$ and $\dot{\mathbf{q}}^0$. The constraint violations result from accumulated integration errors and become more apparent with stiff systems, that is, when the natural frequencies of the system are widely spread. Stiffness can be produced by the physical characteristics of the MBS, such as components with large differences in their masses, stiffness and/or damping. Even though the initial conditions guarantee the nonviolation of constraint equations on position (2.9) and velocity level (2.10), during the course of numerical integration the numerical errors do not satisfy the constraint equations. The effect of these errors increases with time. Hence the constant distances cease to be constant and the points of the same element progressively move closer to or further away from their original position.

When (2.22) is solved, actually (2.11) and (2.12) are solved implicitly. In fact, only the second derivatives of constraint equations will be satisfied in every integration step. Yet it is known that (2.11) represents an unstable system (Franklin et al. 2002). The Baumgarte stabilization method (BSM) allows constraints to be slightly violated before corrective actions can take place, in order to force the violation to vanish. The BSM replaces the differential equation (2.11) by (Baumgarte 1972)

$$\ddot{\mathbf{\Phi}} + 2\alpha\dot{\mathbf{\Phi}} + \beta^2\mathbf{\Phi} = \mathbf{0} \tag{2.23}$$

Equation (2.23) is the differential equation for a closed-loop system in terms of kinematic constraint equations. The terms $2\alpha\dot{\mathbf{\Phi}}$ and $\beta^2\mathbf{\Phi}$ in (2.23) play the role of a control term. The principle of the method is based on the damping of acceleration of constraint violation by feeding back the position and velocity of constraint violations, as seen in Fig. 2.7, which shows open-loop and closed-loop control systems. In the open-loop systems, $\mathbf{\Phi}$ and $\dot{\mathbf{\Phi}}$ do not converge to zero if any perturbation occurs and, therefore, the system is unstable. Thus, utilizing the Baumgarte's approach, the equations of motion for a dynamic system subjected to holonomic constraints are expressed in the form

$$\begin{bmatrix} \mathbf{M} & \mathbf{\Phi}_q^T \\ \mathbf{\Phi}_q & \mathbf{0} \end{bmatrix} \begin{Bmatrix} \ddot{\mathbf{q}} \\ \lambda \end{Bmatrix} = \begin{Bmatrix} \mathbf{g} \\ \gamma - 2\alpha\dot{\mathbf{\Phi}} - \beta^2\mathbf{\Phi} \end{Bmatrix} \tag{2.24}$$

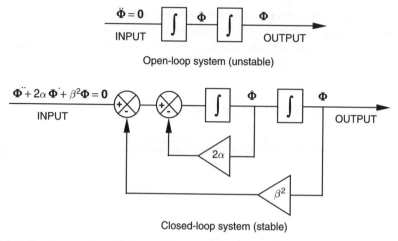

Fig. 2.7 Open-loop (cascade) and closed-loop (feedback) control systems

In general, if α and β are chosen as positive constants, the stability of the general solution of (2.24) is guaranteed. When α is equal to β, critical damping is achieved, which usually stabilizes the system response quickly. Baumgarte (1972) highlighted that the suitable choice of the parameters α and β is performed by numerical experiments. Consequently Baumgarte's method still has some ambiguity in determining optimal feedback gains. Indeed it seems that the value of the parameters is purely empiric, and there is no reliable method for selecting the coefficients α and β. The improper choice of these coefficients can lead to unacceptable results in the dynamic simulation of multibody systems.

Although Baumgarte's constraint stabilization gives good results in most of the applications, it does not help for some particular configuration of the systems, such as near kinematic singularities. Since Baumgarte's approach suffers from the lack of a criterion in the choice of α and β parameters, the fundamental question is: what are the adequate values of α and β parameters for the type of problems involved in this work? Baumgarte (1972) pointed out that the stabilizing values of $\alpha=\beta=5$ is a good choice for a MBS made of rigid bodies. In a later paper, Baumgarte proposed a criterion that aids in the selection of the coefficients, which is further discussed by Kim et al. (1990). In this work, the feedback parameters are inversely proportional to the integration time step when a constant step size is used, $\alpha = 1/\Delta t$ and $\beta = \sqrt{2}/\Delta t$, where Δt is the integration step size. The simplicity of this approach works quite well for computer implementation and appears to be efficient. However, this criterion causes numerical instability and produces incorrect results when the step size is too small. This causes the damping terms to dominate the numerical value of (2.23) and makes the system numerically stiff. Furthermore, if a constant integration step size algorithm is used and it is not very large, this proposed approach damps out the constraint violations faster than an arbitrary assigned constant value of α and β. When a variable step and order integration algorithm is employed, this approach is more likely to lead to numerical problems.

More recently, some works have suggested criteria that help in the selection of the parameters α and β. Chang and Nikravesh (1985) presented an adaptive numerical scheme for determining these parameters. The stability requirement of the adaptation mechanism proposed by Chang and Nikravesh is based on the critical damping, $\alpha=\beta$. The idea of this approach is based on the adaptive scheme in which the feedback parameters are adjusted according to the predicted error. Yet this approach requires complex programming and increases computational time. Later Kim et al. (1990) presented a method for determining the feedback parameters. This method is based on the stability analysis for the error propagation of the modified constrained equation containing the feedback parameters, which is discretized by the Runge–Kutta algorithm in a time domain. Lin and Hong (1998) showed that the parameters α and β can depend on the integration method. In their study, the stability analysis methods used in digital control theory are applied to solve the instability problem.

In short, there is no clear-cut method for choosing the most correct values for feedback parameters for general cases. The choice of α and β parameters usually involves a trial-and-error procedure. Furthermore the selection of the stabilization parameters depends on the type of application, namely, the integrator, the integration step size and the type of model of multibody system. However, the parameters α and β should be, in general, equal to one another and typical values of the stabilization parameters range from 1 to 20 (Nikravesh 1988, Garcia de Jálon and Bayo 1994).

2.7.2 Coordinate Partitioning Method

Based on the original work by Wehage and Haug (1982), the coordinate partitioning method eliminates the errors of the velocities and positions that would otherwise accumulate during the integration process, i.e., the method reduces such errors to values below a specified tolerance. This method requires that sets of independent and dependent coordinates are first identified. Then, only the independent accelerations and velocities are integrated and the dependent quantities are calculated using partitions of the velocity and constraint equations.

The definition of the dependent and independent coordinates can be done automatically by using a matrix factorization technique, such as Gaussian elimination with full pivoting. For a multibody system with m constraints and n coordinates the Jacobian matrix is $m \times n$ and the order of the columns of the matrix corresponds to the order of elements of vector $\mathbf{q}$. Consider a $m \times n$ matrix $\mathbf{A}$ for which the factorization results in the form

$$\mathbf{A} \rightarrow \begin{array}{cc} \overset{m-k}{} & \overset{n-(m-k)}{} \\ \left[\begin{array}{cc} \mathbf{B} & \mathbf{R} \\ \mathbf{S} & \mathbf{D} \end{array} \right] & \begin{array}{c} \}m-k \\ \}k \end{array} \end{array} \qquad (2.25)$$

Assume that there are k redundant rows in the matrix and $m - k$ dependent constraints. As a result of the full pivoting procedure used, the k redundant constraints end up in the factorized matrix bottom rows. $\mathbf{B}$ is a nonsingular $(m - k) \times (m - k)$

matrix associated to the dependent coordinates, and $\mathbf{R}$ is the sub-matrix $(m - k) \times (n - m + k)$ associated to the independent coordinates.

In what follows, let it be assumed that $\mathbf{A}$ represents the Jacobian matrix $\mathbf{\Phi_q}$. Without loss of generality, let it be assumed that there are no redundant constraints in the multibody model or that these have been identified and eliminated. In such case, the Jacobian matrix can be partitioned into

$$\mathbf{\Phi_q} = [\mathbf{\Phi_u} \quad \mathbf{\Phi_v}] \tag{2.26}$$

The Jacobian matrix in (2.26) has the same form of the sub-matrix $[\mathbf{B} \ \mathbf{R}]$ of (2.25). Equation (2.26) implies the partition of the coordinate vector into vectors of dependent and independent coordinates, denoted by $\mathbf{u}$ and $\mathbf{v}$, respectively. This coordinate partition also implies the partition of the velocity and acceleration vectors given as $\dot{\mathbf{q}} = [\dot{\mathbf{u}}^T \dot{\mathbf{v}}^T]^T$ and $\ddot{\mathbf{q}} = [\ddot{\mathbf{u}}^T \ddot{\mathbf{v}}^T]^T$, respectively.

Let it be assumed that the vector of system accelerations is calculated by using (2.15) or (2.18) alternatively. The integration vector $\dot{\mathbf{y}}$, appearing in Fig. 2.6, including only the independent coordinates is

$$\dot{\mathbf{y}} = \begin{bmatrix} \dot{\mathbf{v}}^T & \ddot{\mathbf{v}}^T \end{bmatrix}^T \tag{2.27}$$

which after integration results in vector $\mathbf{y} = \begin{bmatrix} \mathbf{v}^T & \dot{\mathbf{v}}^T \end{bmatrix}^T$.

The dependent velocities and positions are calculated using the velocity and position constraint equations, respectively. To evaluate the dependent velocities let (2.10) be partitioned:

$$\mathbf{\Phi_u}\dot{\mathbf{u}} + \mathbf{\Phi_v}\dot{\mathbf{v}} = \mathbf{\nu} \tag{2.28}$$

The dependent velocities $\dot{\mathbf{u}}$ are evaluated by solving the system of equations:

$$\mathbf{\Phi_u}\dot{\mathbf{u}} = -\mathbf{\Phi_v}\dot{\mathbf{v}} + \mathbf{\nu} \tag{2.29}$$

The dependent positions are obtained through the solution of the position constraint equations; given the independent coordinates, this is

$$\mathbf{\Phi}(\mathbf{u}, \mathbf{v}, t) = \mathbf{0} \tag{2.30}$$

The Newton–Raphson method is used for the solution of (2.30) to obtain the dependent positions $\mathbf{u}$. To achieve convergence some reliable estimates must be provided for these coordinates. A good estimation of $\mathbf{u}^i$ at time t^i, to start the iterative solution procedure, is found by using the information from the previous time t^{i-1} as (Nikravesh 1988)

$$\mathbf{u}^i = \mathbf{u}^{i-1} + h\,\dot{\mathbf{u}}^{i-1} + 0.5h^2\ddot{\mathbf{u}}^{i-1} \tag{2.31}$$

where h is the integration time step from t^i to t^{i-1}.

Summarizing, the methodology that corrects the constraint violations is the coordinate partition method (Wehage and Haug 1982), in which the generalized coordinates are partitioned into independent and dependent sets. The numerical integration is carried out for independent generalized accelerations and independent velocities only leading to the new independent velocities and independent coordinates, respectively. Then the constraint equations are solved for dependent generalized coordinates using, for instance, the Newton–Raphson method. The advantage of this method is that it satisfies all the constraints to the level of precision specified and maintains good error control. However, it suffers from poor numerical efficiency due to the requirement for the iterative solution for dependent generalized coordinates in the Newton–Raphson method, and the constraint velocity equations are solved for the dependent velocities. During integration, numerical problems may arise due to inadequate choice of independent and dependent coordinates that lead to poorly conditioned matrices. For details on this methodology, the interested reader is referred to the work by Wehage and Haug (1982) or Nikravesh (1984, 1988).

2.7.3 Augmented Lagrangian Formulation

The augmented Lagrangian formulation is a methodology that penalizes the constraint violations, much in the same form as the Baumgarte stabilization method (Baumgarte 1972). This is an iterative procedure that presents a number of advantages relative to other methods because it involves the solution of a smaller set of equations, handles redundant constraints and still delivers accurate results in the vicinity of singular configurations.

The augmented Lagrangian formulation consists in solving the system equations of motion, represented by (2.15), by an iterative process. Let index i denote the ith iteration. The evaluation of the system accelerations in a given time step starts as

$$\mathbf{M}\ddot{\mathbf{q}}_i^* = \mathbf{g} \qquad (i = 0) \tag{2.32}$$

The iterative process to evaluate the system accelerations proceeds with the evaluation of

$$\bar{\mathbf{M}}\ddot{\mathbf{q}}_{i+1}^* = \bar{\mathbf{g}} \tag{2.33}$$

where the generalized mass matrix $\bar{\mathbf{M}}$ and load vector $\bar{\mathbf{g}}$ are given by

$$\overline{\mathbf{M}} = \mathbf{M} + \mathbf{\Phi}_{\mathbf{q}}^T \alpha \mathbf{\Phi}_{\mathbf{q}}$$
$$\bar{\mathbf{g}} = \mathbf{M}\ddot{\mathbf{q}}_i^* + \mathbf{\Phi}_{\mathbf{q}}^T \alpha(\gamma_i - 2\omega\mu\mathbf{\Phi}_{\mathbf{q}}\dot{\mathbf{q}}_i^* - \omega^2\mathbf{\Phi}_i) \tag{2.34}$$

In (2.34), the mass matrix $\mathbf{M}$, the Jacobian matrix $\mathbf{\Phi}_{\mathbf{q}}$ and the right-hand side of the acceleration equations γ are the same as those used in (2.15). The penalty terms α, μ and ω ensure that the constraint violation feedbacks are accounted for during the solution of the system equations. The iterative process continues until

$$\left\| \ddot{\mathbf{q}}^*_{i+1} - \ddot{\mathbf{q}}^*_i \right\| \le \varepsilon \tag{2.35}$$

The augmented Lagrangian formulation involves the solution of a system of equations with a dimension equal to the number of coordinates of the multibody system. Though mass matrix $\mathbf{M}$ is generally positive semi-definite the leading matrix of (2.33), $\mathbf{M} + \mathbf{\Phi}^T_\mathbf{q} \alpha \mathbf{\Phi}_\mathbf{q}$, is always positive definite (Garcia de Jálon and Bayo 1994). Even when the system is close to a singular position or when in the presence of redundant constraints the system of equations can still be solved.

2.8 Application Example: Four-Bar Mechanism

An elementary four-bar mechanism is used as a numerical example, to illustrate the influence of the values of Baumgarte's parameters on the control of constraint violations. Figure 2.8 shows the configuration of the four-bar mechanism, which consists of four rigid bodies, including ground, and four revolute joints. The body numbers and their corresponding coordinate systems are shown in Fig. 2.8.

The kinematic joints considered here are assumed to be ideal or perfect, that is, clearance and friction effects are neglected. In the dynamic simulation, the crank is given an initial angular speed of $20\pi\, rad/s$ counter clockwise. The properties of the four-bar mechanism analyzed here are listed in Table 2.1. The numerical parameters concerning the simulation of the four-bar mechanism are listed in Table 2.2.

Several representative simulations are performed to study and compare the efficiency of different values for α and β parameters on the stabilization of the constraint violations. A measure of their efficiency can be drawn from the error of the third constraint equation ($\mathbf{\Phi}_3$) and its derivative ($\dot{\mathbf{\Phi}}_3$) that are representative of the rest of the constraint equations and their derivatives, which present similar results.

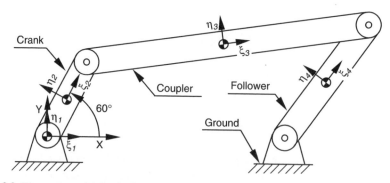

Fig. 2.8 Kinematic model for the four-bar mechanism

Table 2.1 Properties for the individual components of the four-bar mechanism

Body nr	Length (m)	Mass (kg)	Moment of inertia (kg m^2)
1	0.25	–	–
2	0.20	1.00	0.30
3	0.40	2.25	2.00
4	0.40	2.00	1.35

Table 2.2 Simulation parameters for the four-bar mechanism

Crank velocity	20π rad/s
Initial time	0.00 s
Final time	5.00 s
Time step size	0.0001 s

In this case, the third constraint is associated with the revolute joint, connecting crank and coupler.

In the results now presented, only a fixed integration time-step size is used for the dynamic simulation of the four-bar mechanism. The Runge–Kutta–Vener fifth- and sixth-order method, referred to as IVPRK/DIVPRK in the IMSL Fortran Subroutines Library (IMSL 1997), is used to solve the equation of motion with the corresponding initial conditions, as it is not the purpose of the present work to make a detailed description of all available integration algorithms. The interested reader is referred to Brenan et al. (1989) for a more detailed discussion on this issue.

In a first simulation the direct integration method is used, the equations of motion being integrated without any consideration for the constraint violations. Three other simulations are performed considering the stabilization of the constraint violations using the Baumgarte stabilization method. Table 2.3 summarizes the values of the parameters α and β according to the approaches presented and discussed in the previous sections.

Figure 2.9a shows that when direct integration method is used the violation of the constraints grows indefinitely with time. The DIM produced unacceptable results because the constraint equations are rapidly violated due to the inherent instability of the equations used and to the numerical errors that develop during computation. For nonzero values for parameters α and β, the behavior of the multibody system is slightly different, as is illustrated in Fig. 2.9b–d. Moreover, when the parameters α and β are equal, critical damping is reached, which stabilizes the system response more quickly, that is, the first and second derivatives converge to zero. Thus the constraint equations, and not only their second derivatives, are satisfied at any give

Table 2.3 Values of the parameters α and β used in the simulations

Method	α	β
Direct integration method (DIM)	0	0
Baumgarte stabilization method (BSM)	5	5
Baumgarte stabilization method (BSM)	1	23
Baumgarte stabilization method (BSM)	1000	1414

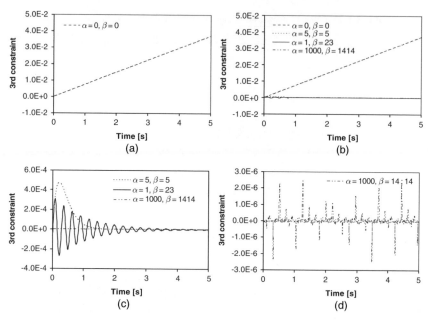

Fig. 2.9 Error of the third constraint equation ($\mathbf{\Phi}_3$)

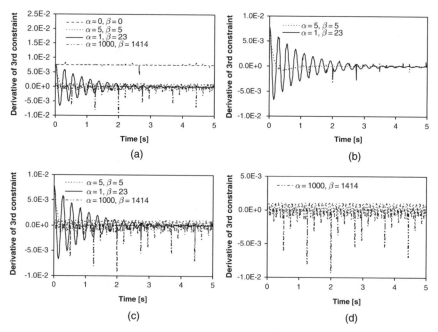

Fig. 2.10 Error of derivative of the third constraint equation ($\dot{\mathbf{\Phi}}_3$)

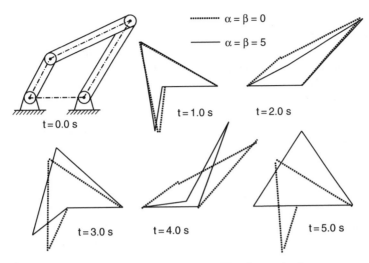

Fig. 2.11 Animation sequence of dynamic simulation of four-bar mechanism

time. Figure 2.9d illustrates a stiff system, which occurs when the values of α and β are high.

Figure 2.10a–d depicts the violation of the derivative of the third constraint equation ($\dot{\Phi}_3$). It should be highlighted that different scales are used for results plotted in Figs. 2.9a–d and 2.10a–d, in order to clearly observe the influence of the values of Baumgarte parameters. Figure 2.11 shows a sequence of configurations that allows the visualization of the constraint violations during the dynamic simulation of the four-bar mechanism.

2.9 Summary

In this chapter, the fundamentals of the formulation for the kinematic and dynamic analyses of multibody systems were presented and discussed. The formulation closely follows the work of Nikravesh (1988) and summarizes how the general equations of motion of dynamic equilibrium of a MBS can be formulated, using generalized Cartesian coordinates and one Newton–Euler method. This formulation is straightforward in terms of assembling equations of motion and providing all reaction forces. From the mathematical viewpoint, Cartesian coordinates are the supporting structure for all the methodologies and dynamic analysis presented in the forthcoming chapters. Furthermore the main aspects related to the formulation of the planar revolute and translational joints were also reviewed.

The equations of motion for constrained MBS are expressed by (2.15), which is often referred to as a coupled set of differential and algebraic equations. A set of initial conditions, positions and velocities, is required to start the dynamic simulation. The selection of the appropriate initial conditions plays a key role in the prediction of the dynamic performance of mechanical systems. The subsequent initial conditions for each time step in the simulation are obtained from the final conditions of the previous time step, i.e., from the initial values for positions and velocities (2.15) is solved for accelerations, $\ddot{\mathbf{q}}$, and Lagrange multipliers, $\boldsymbol{\lambda}$, using the L-U factorization in conjunction with forward and backward substitution. The positions and velocities at the next time step are then obtained by integration of the velocity and acceleration vectors, $\dot{\mathbf{q}}$ and $\ddot{\mathbf{q}}$. This procedure is repeated until the final time is reached. The integration process can be performed using a constant step size scheme, such as the Runge–Kutta method, or a predictor–corrector algorithm with both variable step and order (Shampine and Gordon 1975, Gear 1981). It was not the purpose of this chapter to make a detailed description of all the available methods and integration algorithms. The interested reader is referred to the work by Brenan et al. (1989).

In order to stabilize or keep under control the constraint violations, (2.15) is solved using the Baumgarte stabilization method (Baumgarte 1972). It was shown in this chapter that the numerical instability, during the numerical integration of kinematic acceleration equations, suggests that direct integration method may not be used directly to obtain the solution of the dynamic simulation problem. In order to ensure that the system constraint equations are fulfilled, the Baumgarte stabilization method or any other corrective measure must be implemented. The introduction of feedback gains can eliminate the growth of the constraint violations within a stable range, but there is no clear-cut suggestion for choosing values for feedback parameters for general cases. However, it was demonstrated that for α and β equal to each other and eventually equal to 5, the system response stabilizes more quickly. However, the constraint violations can only be completely eliminated by using the coordinate partition method. The BSM not only can fail in the cases of configuration singularities but also is dependent on the empiric feedback gains. Instead of representing constraints by the unstable differential equation, it was suggested to introduce them in the form of an asymptotically stable differential equation. With this methodology, experience tells that the numerical result does not diverge from the exact solution, but oscillates around it. Magnitude and frequency of the oscillation depend on the values of Baumgarte parameters.

References

Ambrósio J (1991) Elastic-plastic large deformation of flexible multibody systems in crash analysis. Ph.D. Dissertation, University of Arizona, Tucson, AZ.

Baumgarte J (1972) Stabilization of constraints and integrals of motion in dynamical systems. Computer Methods in Applied Mechanics and Engineering 1:1–16.

Bayo E, Garcia de Jálon J, Serna A (1988) A modified Lagrangian formulation for the dynamic analysis of constrained mechanical systems. Computer Methods in Applied Mechanics and Engineering 71:183–195.

Blajer W (1999) Elimination of constraint violation and accuracy improvement in numerical simulation of multibody systems. Proceedings of EUROMECH colloquium 404, Advances in computational multibody dynamics, IDMEC/IST, Lisbon, Portugal, September 20–23, edited by J Ambrósio and W Schiehlen, pp. 769–787.

Brenan KE, Campbell SL, Petzold LR (1989) The numerical solution of initial value problems in differential-algebraic equations. North-Holland, New York.

Chang CO, Nikravesh PE (1985) An adaptive constraint violation stabilization method for dynamic analysis of mechanical systems. Journal of Mechanisms, Transmissions, and Automation in Design 107:488–492.

Franklin GF, Powel JD, Enami-Naeini A (2002) Feedback control of dynamic systems, Fourth edition. Prentice Hall, Englewood Cliffs, NJ.

Garcia de Jálon J, Bayo E (1994) Kinematic and dynamic simulations of multibody systems. Springer, Berlin Heidelberg New York.

Gear CW (1981) Numerical solution of differential-algebraic equations. IEEE Transactions on Circuit Theory CT-18:89–95.

Huston RL (1990) Multibody dynamics. Butterworth-Heinemann, Boston, MA.

IMSL Fortran 90 Math Library 4.0 (1997) Fortran subroutines for mathematical applications. Visual Numerics, Inc., Huston, TX.

Jiménez JM, Avello A, García-Alonso A, Garcia de Jálon J (1990) COMPAMM—a simple and efficient code for kinematic and dynamic numerical simulation of 3-D multibody system with realistic graphics. Multibody systems handbook. Springer, Berlin Heidelberg New York.

Kim JK, Chung IS, Lee BH (1990) Determination of the feedback coefficients for the constraint violation stabilization method. Proceedings of Institution Mechanical Engineers 204:233–242.

Lin S, Hong M (1998) Stabilization method for numerical integration mechanical systems. Journal of Mechanical Design 120:565–572.

Lin S, Huang J (2002) Stabilization of Baumgarte's method using the Runge Kutta approach. Journal of Mechanical Design 124:633–641.

Neto MA, Ambrósio J (2003) Stabilization methods for the integration of DAE in the presence of redundant constraints. Multibody System Dynamics 10:81–105.

Nikravesh PE (1984) Some methods for dynamic analysis of constrained mechanical systems: a survey. Computer-aided analysis and optimization of mechanical system dynamics, edited by EJ Haug, Springer, Berlin Heidelberg New York pp. 351–368.

Nikravesh PE (1988) Computer-aided analysis of mechanical systems. Prentice Hall, Englewood Cliffs, NJ.

Park KC, Chiou JC (1988) Stabilization of computational procedures for constrained dynamical systems. Journal of Guidance, Control, and Dynamics 11:365–370.

Rosen A, Edelstein E (1997) Investigation of a new formulation of the Lagrange method for constrained dynamic systems. Journal of Applied Mechanics 64:116–122.

Rulka W (1990) SIMPACK—a computer program for simulation of large motion multibody systems. Multibody systems handbook, edited by W. Schichlen, pp. 265–284. Springer, Berlin Heidelberg New York.

Ryan RR (1990) ADAMS—multibody system analysis software. Multibody systems handbook. Springer, Berlin Heidelberg New York.

Schiehlen W (1990) Multibody systems handbook. Springer, Berlin Heidelberg New York.

Shabana AA (1989) Dynamics of multibody systems. Wiley, New York.

Shampine L, Gordon M (1975) Computer solution of ordinary differential equations: the initial value problem. Freeman, San Francisco, CA.

Smith RC, Haug EJ (1990) DADS—Dynamic analysis and design system. Multibody systems handbook. Springer, Berlin Heidelberg New York.

Wehage RA, Haug EJ (1982) Generalized coordinate partitioning for dimension reduction in analysis of constrained systems. Journal of Mechanical Design 104:247–255.

Wittenburg J (1977) Dynamics of systems of rigid bodies. B.G. Teubner, Stuttgart, Germany.

Yoon S, Howe RM, Greenwood DT (1994) Geometric elimination of constraint violations in numerical simulation of lagrangian equations. Journal of Mechanical Design 116:1058–1064.

Chapter 3
Contact-Impact Force Models
for Mechanical Systems

The collision is a prominent phenomenon in many mechanical systems that involve intermittent motion, kinematic discontinuities or clearance joints. As a result of an impact, the values of the system state variables change very fast, eventually looking like discontinuities in the system velocities and accelerations. The impact is characterized by large forces that are applied and removed in a short time period. The knowledge of the peak forces developed in the impact process is very important for the dynamic analysis of multibody mechanical systems having consequences in the design process. The numerical description of the collision phenomenon is strongly dependent on the contact-impact force model used to represent the interaction between the system components. The model for the contact-impact force must consider the material and geometric properties of the colliding surfaces and information on relative positions and velocities, contribute to an efficient integration and account for some level of energy dissipation. These characteristics are ensured with a continuous contact force model, in which the deformation and contact forces are considered as continuous functions during the complete period of contact. This chapter deals with contact-impact force models for both spherical and cylindrical contact surfaces. The incorporation of the friction phenomenon, based on the Coulomb's friction law, is also discussed together with a computational strategy which includes an automatic step size selection procedure based not only on numerical error control but also on the characteristics of the contact.

3.1 Approaches to Contact and Impact of Rigid Bodies

Impact occurs during the collision of two or more bodies, which may be external or belong to a multibody mechanical system. The impact phenomenon is characterized by abrupt changes in the values of system variables, most visible as discontinuities in the system velocities. Other effects directly related to the impact phenomena are the vibration propagation on the system components, local elastic/plastic deformations at the contact zone and some level of energy dissipation. The impact is a very important phenomenon in many mechanical systems such as mechanisms with intermittent motion and with clearance joints (Khulief and Shabana 1986, Ravn 1998).

The selection of the most adequate contact force model plays a key role in the correct design and analysis of these types of mechanical systems (Flores et al. 2006).

By and large, an impact may be considered to occur in two phases: the compression or loading phase and the restitution or unloading phase. During the compression phase, the two bodies deform in the normal direction to the impact surfaces, and the relative velocity of the contact points/surfaces on the two bodies in that direction is gradually reduced to zero. The end of the compression phase is referred to as the instant of maximum compression or maximum approach. The restitution phase starts at this point and ends when the two bodies separate from each other (Brach 1991). The restitution coefficient reflects the type of collision. For a fully elastic contact the restitution coefficient is equal to the unit, while for a fully plastic contact restitution coefficient is null. The most general and predominant type of collision is the oblique eccentric collision, which involves both relative normal velocity and relative tangential velocity (Maw et al. 1975, Zukas et al. 1982).

In order to evaluate efficiently the contact-impact forces resulting from collisions in multibody systems, such as the contact between the bearing and journal in a revolute joint with clearance, special attention must be given to the numerical description of the contact force model. Information on the impact velocity, material properties of the colliding bodies and geometry characteristics of the contact surfaces must be included in the contact force model. These characteristics are observed with a continuous contact force, in which the deformation and contact forces are considered as continuous functions (Lankarani and Nikravesh 1990). Furthermore it is important that the contact force model can add to the stable integration of the equation of motion of multibody system.

In a broad sense, there are two different methods to solve the impact problem in multibody mechanical systems designated as continuous and discontinuous approaches (Lankarani and Nikravesh 1990). Within the continuous approach the methods commonly used are the continuous force model, which is in fact a penalty method, and the unilateral constraint methodology, based on the linear complementary approach (Pfeiffer and Glocker 1996). The continuous contact force model represents the forces arising from collisions and assumes that the forces and deformations vary in a continuous manner. In this method, when contact between the bodies is detected, a normal force perpendicular to the plane of collision is applied. This force is typically applied as a spring–damper element, which can be linear, such as the Kelvin–Voigt model (Lankarani 1988), or nonlinear, such as the Hunt and Crossley model (Hunt and Crossley 1975). For long impact durations this method is effective and accurate in that the instantaneous contact forces are introduced into the system's equations of motion. The second continuous methodology specifies that when contact is detected a kinematic constraint is introduced in the system's equations. Such a constraint is maintained while the reaction forces are compressive, and removed when the impacting bodies rebound from contact (Ambrósio 2000).

A second approach of a different nature is a discontinuous method that assumes that the impact occurs instantaneously, the integration of the equations of motion being halted at the time of impact. Then a momentum balance is performed to calculate the post-impact velocity, the integration being resumed afterwards with the updated velocities, until the next impact occurs. In the discontinuous method, the dynamic analysis of the system is divided into two intervals, before and after impact. The restitution coefficient is employed to quantify the dissipation energy during the impact. The restitution coefficient only relates relative velocities after separation to relative velocities before contact and ignores what happens in between. The discontinuous method is relatively efficient but the unknown duration of the impact limits its application, since for large enough contact periods the system configuration changes significantly (Lankarani 1988). Hence the assumption of instantaneity of impact duration is no longer valid and the discontinuous analysis must not be adopted. This method, commonly referred to as piecewise analysis, has been used for solving the intermittent motion problem in mechanical systems (Khulief and Shabana 1986) and it is still the most commonly used approach in vehicle accident reconstruction (PC-Crash 2002).

In general, the contact points change during collision. When there is no penetration between the colliding bodies, there is no contact and, consequently, the contact forces are null. The occurrence of penetration is used as the basis to develop the procedure to evaluate the local deformation of the bodies in contact. Although the bodies are assumed to be rigid, the contact forces correspond to those evaluated as if the penetration is due to local elastic deformations. These forces are calculated as being equivalent to those that would appear if the bodies in contact were pressed against each other by an external static force. This means that the contact forces are treated as elastic forces expressed as functions of the coordinates and velocities of the colliding bodies. The methodology used here allows for the accurate calculation of the location of contact points. The direction of the normal contact force is determined from the normal vector to the plane of colliding surfaces at the points of contact.

In short, in dynamic analysis, the deformation is known at every time step from the configuration of the system and the forces are evaluated based on the state variables. With the variation of the contact force during the contact period, the response of the dynamic system is obtained by simply including updated forces into the equations of motion. Since the equations of motion are integrated over the period of contact, this approach results in a rather accurate response. Furthermore this methodology is not limited by the changes in the system configuration during the contact periods.

3.2 Normal Force Models for Spherical Contact Surfaces

The simplest contact force relationship, known as Kelvin–Voigt viscous-elastic model, is modeled by a parallel spring–damper element (Zukas et al. 1982). The spring represents the elasticity of the contacting bodies while the damper describes the loss of kinetic energy during the impact. In most studies, the stiffness and damping coefficients have been assumed to be known parameters, and the analysis has been confined to unconstrained bodies. The spring stiffness in the element can be

calculated using a simple mechanical formula or obtained by means of the finite
element method (FEM). Recently Zhu et al. (1999) proposed a theoretical formula
for calculating damping in the impact of two bodies in a multibody system. This
model assumes that both the spring and the damper are linear. When the contact
bodies are separating from each other the energy loss is included in the contact
model by multiplying the rebound force with a coefficient of restitution. The restitution coefficient accounts for the energy dissipated during the impact in the form
of a hysteresis in the relation between force and deformation.

The normal Kelvin–Voigt contact force, F_N, is calculated for a given penetration
depth, δ, as

$$F_N = \begin{cases} K\delta & \text{if } v_N > 0 \quad \text{(loading phase)} \\ K\delta\,c_e & \text{if } v_N < 0 \quad \text{(unloading phase)} \end{cases} \qquad (3.1)$$

where K is the stiffness, δ is the relative penetration depth, c_e is the restitution
coefficient and v_N is the relative normal velocity of the colliding bodies.

The primary drawback associated with this model is the quantification of the
spring constant, which depends on the geometry and material properties of the contacting bodies. On the other hand, the assumption of a linear relation between the
penetration depth and the contact forces is at best a rough approximation because
the contact force depends on the shape, surface conditions and material properties
of the contacting surfaces, all of which suggest a more complex relation.

For the linear Kelvin–Voigt model, Fig. 3.1a–c shows the penetration depth δ,
the normal contact force F_N and the hysteresis of two internally colliding spheres.
The restitution coefficient and the spring stiffness used to build Fig. 3.1 are 0.9 and
$1.5 \times 10^8 \, N/m$, respectively.

The best-known contact force law between two spheres of isotropic materials is
due to the result of the work by Hertz, which is based on the theory of elasticity
(Timoshenko and Goodier 1970). The Hertz (1896) contact theory is restricted to
frictionless surfaces and perfectly elastic solids being exemplified by the case shown
in Fig. 3.2.

The Hertz law relates the contact force with a nonlinear power function of penetration depth and is written as

$$F_N = K\delta^n \qquad (3.2)$$

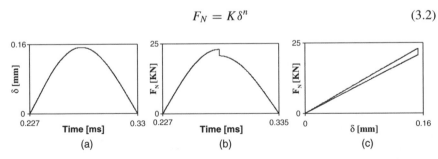

Fig. 3.1 Internally colliding spheres modeled by linear Kelvin–Voigt viscous-elastic contact
model: (**a**) penetration depth, d; (**b**) normal contact force, F_N; (**c**) force–penetration relation

Fig. 3.2 Relative penetration depth during the impact between two spheres

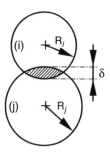

where K is the generalized stiffness constant and δ is the relative normal indentation between the spheres. The exponent n is set to 1.5 for the cases where there is a parabolic distribution of contact stresses, as in the original work by Hertz. Although for metallic materials $n = 1.5$ for other materials such as glass or polymer it can be either higher or lower, leading to a convenient contact force expression that is based on experimental work but that should not be confused with Hertz theory. The generalized parameter K is dependent on the material properties and the shape of the contact surfaces. For two spheres in contact the generalized stiffness coefficient is a function of the radii of the spheres i and j and the material properties as (Goldsmith 1960)

$$K = \frac{4}{3(\sigma_i + \sigma_j)} \left[\frac{R_i R_j}{R_i + R_j} \right]^{\frac{1}{2}} \tag{3.3}$$

where the material parameters σ_i and σ_j are given by

$$\sigma_k = \frac{1 - \nu_k^2}{E_k}, \quad (k = i, j) \tag{3.4}$$

and the quantities ν_k and E_k are the Poisson's ratio and the Young's modulus associated with each sphere, respectively. For contact between a sphere body i and a plane surface body j the generalized stiffness coefficient depends on the radius of the sphere and the material properties of the contacting surfaces, being expressed by (Goldsmith 1960)

$$K = \frac{4}{3(\sigma_i + \sigma_j)} \sqrt{R_i}. \tag{3.5}$$

For two internally colliding spheres modeled by Hertz contact law, Fig. 3.3a–c shows the penetration depth, δ, the normal contact force, F_N, and the relation force–penetration. The generalized stiffness is equal to $6.6 \times 10^{10} \, N/m^{1.5}$ for the calculations used to generate the graphs.

It is apparent that the Hertz contact law given by (3.2) cannot be used during both phases of contact (loading and unloading phases), since this model does not take into account the energy dissipation during the process of impact. This is a pure elastic contact model, that is, the contact energy stored during the loading phase

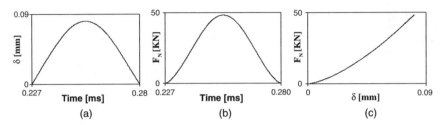

Fig. 3.3 Internally colliding spheres modeled by Hertz contact law: (**a**) penetration depth, δ; (**b**) normal contact force, F_N; (**c**) force–penetration ratio

is exactly the same that is restored during the unloading phase. The advantages of the Hertz law relative to Kelvin–Voigt contact model reside on its physical meaning represented by both its nonlinearity and by the relation between the generalized stiffness and geometry and material of the contacting surfaces. Although the Hertz law is based on the elasticity theory, some studies have been performed to extend its application to include energy dissipation. In fact, the process of energy transfer is a complicated part of modeling impacts. If an elastic body is subjected to a cyclic load, the energy loss due to internal damping causes a hysteresis loop in the force–displacement diagram, which corresponds to energy dissipation.

Hunt and Crossley (1975) showed that the linear spring–damper model does not represent the physical nature of energy transferred during the impact process. Instead they represent the contact force by the Hertz force law with a nonlinear viscous-elastic element. This approach is valid for direct central and frictionless impacts. Based on Hunt and Crossley's work, Lankarani and Nikravesh (1990) developed a contact force model with hysteresis damping for impact in multibody systems. The model uses the general trend of the Hertz law, the hysteresis damping function being incorporated to represent the energy dissipated during the impact. Lankarani and Nikravesh (1990) suggested separating the normal contact force into elastic and dissipative components as

$$F_N = K\delta^n + D\dot{\delta} \tag{3.6}$$

where the first term of the right-hand side is referred to as the elastic force and the second term accounts for the energy dissipated during the impact. In (3.6), the quantity D is a hysteresis coefficient and $\dot{\delta}$ is the relative normal impact velocity.

The hysteresis coefficient is written as a function of penetration as

$$D = \chi \, \delta^n \tag{3.7}$$

in which the hysteresis factor χ is given by

$$\chi = \frac{3K(1 - c_e^2)}{4\dot{\delta}^{(-)}} \tag{3.8}$$

$\delta^{(-)}$ being the initial impact velocity. By substituting (3.8) into (3.7) and the results into (3.6), the normal contact force is finally expressed as

$$F_N = K\delta^n \left[1 + \frac{3(1 - c_e^2)}{4} \frac{\dot{\delta}}{\dot{\delta}^{(-)}} \right] \tag{3.9}$$

where the generalized parameter K is evaluated by (3.3) and (3.4) for sphere-to-sphere contact or by similar expressions for the contact of other types of geometry; c_e is the restitution coefficient, $\dot{\delta}$ is the relative normal penetration velocity and $\dot{\delta}^{(-)}$ is the initial normal impact velocity where contact is detected. The use of the damping scheme included in this model implies the outcome illustrated in Fig. 3.4a–c in which the penetration depth, δ, normal contact force, F_N, and hysteresis of an impact between two internally colliding spheres are presented. The generalized stiffness used to evaluate the relations in Fig. 3.4 is $6.6 \times 10^{10}\,\mathrm{N}/m^{1.5}$.

Equation (3.9) is valid only for impact velocities lower than the propagation velocity of elastic waves across the bodies, that is, $\dot{\delta}^{(-)} \le 10^{-5}\sqrt{E/\rho}$, where E is the Young's modulus and ρ is the material mass density (Love 1944). The quantity $\sqrt{E/\rho}$, velocity of wave propagation, is the larger of two propagation velocities of the elastic deformation waves in the colliding bodies.

Shivaswamy (1997) studied theoretically and experimentally the impact between bodies and demonstrated that at low impact velocities, the hysteresis damping is the prime factor for energy dissipation. Impact at higher velocities, exceeding the propagation velocity of the elastic deformation waves, is likely to dissipate energy in a form not predicted by the current model. In a later work, Lankarani and Nikravesh (1994) proposed a new approach for contact force analysis, in which the permanent indentation is also included. At fairly moderate or high velocities of collision, especially in the case of metallic solids, permanent indentations are left behind on the colliding surfaces. Hence local plasticity of the surfaces in contact becomes the dominant source of energy dissipation during impact. Permanent or plastic deformations are beyond the scope of the present work.

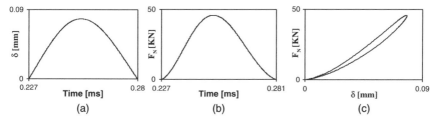

Fig. 3.4 Internally colliding spheres modeled by the Lankarani and Nikravesh contact force model: (a) penetration depth, δ; (b) normal contact force, F_N; (c) force–penetration relation

3.3 Normal Contact Force Models for Cylindrical Surfaces

The contact models given by (3.2) and (3.9) are valid only for colliding bodies that exhibit in the contacting surface a parabolic contact stress distribution, such as in the case of ellipsoidal contact areas. For a cylindrical contact area such as the one pictured in Fig. 3.5 between two parallel cylinders, a literature search reveals few approximate force–displacement relationships.

It is worth noting that line contact assumes a precise parallel alignment of the colliding cylinders. Furthermore a uniform force distribution over the length of the cylinders is also assumed and boundary effects are neglected. For the case of cylindrical contact forces, some authors suggest the use of the more general and straightforward force–displacement relation given by (3.9) but with an exponent, n, in the range of 1–1.5 (Hunt and Crossley 1975, Ravn 1998). Dietl (1997) used the classical solution of contact, presented by Hertz, but with the exponent n equal to 1.08 to model the contact between the journal and the bearing elements.

Based on Hertz theory, Dubowsky and Freudenstein (1971) proposed an expression for the indentation, as function of the contact force, of an internal pin inside a cylinder as

$$\delta = F_N \left(\frac{\sigma_i + \sigma_j}{L} \right) \left[\ln \left(\frac{L^b (R_i - R_j)}{F_N R_i R_j (\sigma_i + \sigma_j)} \right) + 1 \right] \tag{3.10}$$

where $R_{i,j}$ and $\sigma_{i,j}$ are the parameters shown in (3.4), L is the length of the cylinder and the exponent b has a value 3. Since (3.10) is a nonlinear implicit function for F_N, with a known penetration depth, F_N can be evaluated. This is a nonlinear problem and requires an iterative solution scheme, such as the Newton–Raphson method, to solve for the normal contact force, F_N.

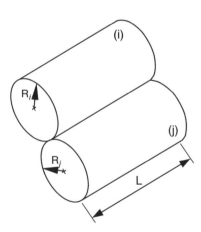

Fig. 3.5 Contact between two external cylinders

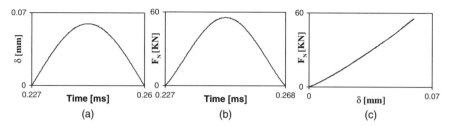

Fig. 3.6 Internally colliding cylinders using Dubowsky and Freudenstein contact force model: (**a**) penetration depth, δ; (**b**) normal contact force, F_N; (**c**) force–penetration ratio

Based on the Dubowsky and Freudenstein contact force model the solutions corresponding to the time variation of the indentation δ, the normal contact force F_N and the force–penetration depth ratio are shown in Fig. 3.6a–c. In the plots presented for the case of a pin contact inside a cylinder, the pin and cylinder radii are 9.5 and 10 mm, respectively. The length of the cylinder is equal to 15 mm, and both the pin and the cylinder are made of steel.

Goldsmith (1960) proposed an expression similar to (3.10) but with the value of exponent b equal to 1. However, this value for b leads to a problem of consistency of the units in the expression. Figure 3.7a–c shows the penetration depth δ, the normal contact force F_N and the force–penetration ratio of two internally colliding cylinders modeled with the Goldsmith contact force. This model shows that the force–penetration ratio is almost linear.

The ESDU 78035 Tribology Series (1978) also proposes some expressions for contact mechanics analysis suitable for engineering applications. For a circular contact area the ESDU 78035 model is the same as the pure Hertz law given by (3.2). For rectangular contact, e.g., a pin inside a cylinder, the expression is given by

$$\delta = F_N \left(\frac{\sigma_i + \sigma_j}{L} \right) \left[\ln \left(\frac{4L(R_i - R_j)}{F_N(\sigma_i + \sigma_j)} \right) + 1 \right] \tag{3.11}$$

where all quantities are the same as used for the calculations in Fig. 3.6.

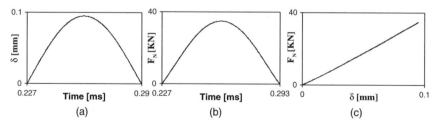

Fig. 3.7 Internally colliding cylinders using Goldsmith contact force model: (**a**) penetration depth, δ; (**b**) normal contact force, F_N; (**c**) force–penetration ratio

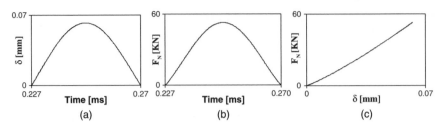

Fig. 3.8 Internally colliding cylinders using the ESDU 78035 contact force model: (**a**) penetration depth, δ; (**b**) normal contact force, F_N; (**c**) force–penetration ratio

Figure 3.8a–c shows the penetration depth δ, the normal contact force F_N and the force–penetration ratio of two internally colliding cylinders modeled by ESDU 78035, given by (3.11).

The contact force due to the spherical and cylindrical contact force models is displayed in Fig. 3.9, where it can be observed that the spherical and cylindrical force–displacement relations are reasonably close. Thus the straightforward force–penetration relation proposed by Lankarani and Nikravesh given in (3.9) is largely used for mechanical contacts not only because of its simplicity and easiness in implementation in a computational program, but also because this is the only model that accounts for the energy dissipation during the impact process (Ryan 1990, Smith and Haug 1990, Bottasso et al. 1999, Pedersen 2001, Pedersen et al. 2002, Silva and Ambrósio 2004, Flores et al. 2006).

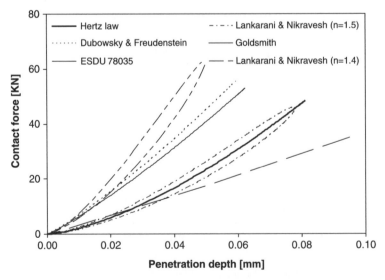

Fig. 3.9 Force–deformation curves for spherical and cylindrical contact surfaces

3.4 Tangential Friction Force Models

When contacting bodies slide or tend to slide relative to each other tangential forces are generated. These forces are usually referred to as friction forces. Three basic principles have been experimentally established, namely: (1) the friction force acts in a direction opposite to that of the relative motion between the two contacting bodies; (2) the friction force is proportional to the normal load on the contact; (3) the friction force is independent of a nominal area of contact. These three statements constitute what is known as the laws of sliding friction under dry conditions (Stolarski 1990). Based on the experimental observations by Angmonds in the field of optics, Coulomb developed what is today the most commonly used friction law, but there is still no simple model which can be universally used by designers to calculate the friction force for a given pair of bodies in contact. In face of the shortcomings of the Coulomb friction law, in recent years there has been much interest on the subject of friction, and many research papers have focused on the subject (Keller 1986, Wang and Manson 1992, Han and Gilmore 1993).

The presence of friction in the contact surfaces makes the contact problem more complicated as the friction may lead to different contact modes, such as sticking or sliding. For instance, when the relative tangential velocity of two impacting bodies approaches zero, stiction occurs. Indeed, as pointed out by Ahmed et al. (1999), the friction model must be capable of detecting sliding, sticking and reverse sliding to avoid energy gains during impact. This work was developed for the treatment of impact problems in jointed open-loop multibody systems. Lankarani (2000) extends Ahmed's formulation to the analysis of impact problems with friction in any general multibody system including both open- and closed-loop systems.

The Coulomb's friction law of sliding friction can represent the most fundamental and simplest model of friction between dry contacting surfaces. When sliding takes place, the Coulomb law states that the tangential friction force F_T is proportional to the magnitude of the normal contact force, F_N, at the contact point by introducing a coefficient of friction c_f (Greenwood 1965). The Coulomb's friction law is independent of relative tangential velocity. In practice, this is not true because friction forces can depend on many parameters such as material properties, temperature, surface cleanliness and velocity of sliding, which cannot all be accumulated for by a constant friction coefficient. Therefore a continuous friction force–velocity relationship is desirable. Furthermore the application of the original Coulomb's friction law in a general-purpose computational program may lead to numerical difficulties because it is a highly nonlinear phenomenon that may involve switching between sliding and stiction conditions. Also from this point of view, more realistic friction force models are required.

In the last decades, a number of papers addressed the issue of the tangential friction forces (Bagci 1975, Threlfall 1978, Rooney and Deravi 1982, Haug et al. 1986, Wu et al. 1986a, b). Most of them use the Coulomb friction model with some modification in order to avoid the discontinuity at zero relative tangential velocity and to obtain a continuous friction force. Dubowsky (1974) assumed the friction force to be equal to a constant value opposing the direction of velocity, given by

$$\mathbf{f}_T = -c_c \frac{\mathbf{v}_T}{v_T} \tag{3.12}$$

where c_c is a coefficient independent of normal contact force and $\mathbf{v}_T$ is the relative tangential velocity. This model does not take the effect of zero velocity into account, that is, it has the disadvantage of an infinite gradient at null relative tangential velocity. This causes computational difficulties in the integration process since the force instantaneously changes from $-\mathbf{f}_T$ to $+\mathbf{f}_T$. This model is qualitatively illustrated in Fig. 3.10a, which shows the Coulomb's friction force versus relative tangential velocity.

A friction model with better numerical features is found in Rooney and Deravi (1982), where the friction force is calculated from two sets of equations. When the relative tangential velocity is not close to zero the Coulomb's friction law is given by

$$\mathbf{f}_T = -c_f \mathbf{f}_N \frac{\mathbf{v}_T}{v_T} \tag{3.13}$$

and when the relative tangential velocity of the contacting bodies is close to zero the friction force is a value within a range given by

$$-c_f F_N < F_T < c_f F_N \tag{3.14}$$

where c_f is the coefficient of friction, $\mathbf{v}_T$ is the relative tangential velocity and F_N is the normal contact force, which is always positive. This model is illustrated in Fig. 3.10b.

Wilson and Fawcett (1974) used the model expressed by (3.13) in the dynamic study of a slider–crank mechanism with a clearance joint between the slider and the guide. More recently, Ravn (1998) also used (3.13) to include the friction effect in revolute joints with clearance. Threlfall (1978) proposed another friction force model, in which the transition between $-\mathbf{f}_T$ and $+\mathbf{f}_T$ is made using a curve as follows:

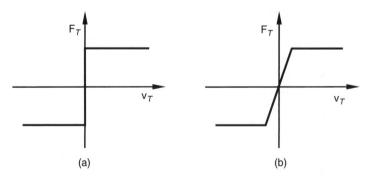

(a)   (b)

Fig. 3.10 (a) Standard Coulomb's friction law, (b) Rooney and Deravi friction force

$$\mathbf{f}_T = c_f \mathbf{f}_N \frac{\mathbf{v}_T}{v_T} \left(1 - e^{-\left(\frac{3v_T}{v_r}\right)}\right), \quad \text{if } |\mathbf{v}_T| < v_r \tag{3.15}$$

where c_f is the friction coefficient, $\mathbf{f}_N$ is the normal contact force, v_r is a small characteristic velocity as compared to the maximum relative tangential velocity encountered during the simulation. The value of v_r is a specified parameter that for small values results in slowing down the integration method as it gets closer to the idealized model of Fig. 3.10a. In practice, the regulation factor $1 - \exp(-3v_T/v_r)$ smoothes out the friction force discontinuity. The shape of this curve is illustrated in Fig. 3.11a.

Ambrósio (2002) presented another modification for Coulomb's friction law, in which the dynamic friction force is expressed as

$$\mathbf{f}_T = -c_f c_d \mathbf{f}_N \frac{\mathbf{v}_T}{v_T} \tag{3.16}$$

where c_f is the friction coefficient, F_N is the normal contact force, $\mathbf{v}_T$ is the relative tangential velocity and c_d is a dynamic correction coefficient, which is expressed as

$$c_d = \begin{cases} 0 & \text{if} & v_T \le v_0 \\ \frac{v_T - v_0}{v_1 - v_0} & \text{if} & v_0 \le v_T \le v_1 \\ 1 & \text{if} & v_T \ge v_1 \end{cases} \tag{3.17}$$

in which v_0 and v_1 are given tolerances for the velocity. This dynamic correction factor prevents the friction force from changing direction for almost null values of the tangential velocity, which is perceived by the integration algorithm as a dynamic response with high-frequency contents, thereby forcing a reduction in the time-step size. The great merit of this modified Coulomb's law is that it allows the numerical stabilization of the integration algorithm. This friction force model, illustrated in Fig. 3.11b, does not account for other phenomena like the adherence between the sliding contact surfaces, which can be added as a complementary model.

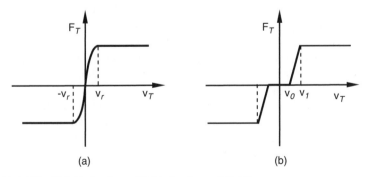

Fig. 3.11 (a) Threlfall friction force, (b) friction force of (3.16)

3.5 Numerical Aspects in Contact Analysis

In a dynamic simulation, it is very important to find the precise instant of transition between the different states, that is, contact and noncontact situations. This requires close interaction with the numerical procedure to continuously detect and analyze all situations. If not, errors may build up and the final results will become inaccurate.

When a system consists of fast- and slow-moving components, that is, the *eigenvalues* are widely spread, the system is designated as being stiff (Nikravesh 1988). Stiffness in the system equations of motion arises when the gross motion of the overall mechanism is combined with the nonlinear contact forces that lead to rapid changes in velocity and accelerations. In addition, when the equations of motion are described by a coupled set of differential and algebraic equations, the error of the response system is particularly sensitive to constraints violation. Constraints violation inevitably leads to artificial and undesired changes in the energy of the system. Yet, by applying a stabilization technique, the constraint violation can be kept under control (Baumgarte 1972). During the numerical integration procedure, both the order and the step size are adjusted to keep the error tolerance under control. In particular, the variable step size of the integration scheme is a desirable feature when integrating systems that exhibit different time scales, such as in multibody systems with impacting bodies (Shampine and Gordon 1975). Thus large steps are taken when the system's motion does not include contact forces, and when impact occurs the step size is decreased substantially to capture the high-frequency response of the system.

One of the most critical aspects in the dynamic simulation of the multibody systems with collisions is the detection of the precise instant of contact. In addition, the numerical model used to characterize the contact between the bodies requires the knowledge of the pre-impact conditions, that is, the impact velocity and the direction of the plane of collision. The contact duration and the penetration cannot be predicted from the pre-impact conditions due to the influence of the kinematic constraints imposed in the bodies on the overall system motion. Thus, before the first impact, the bodies can freely move relative to each other and, in this phase, the step size of the integration algorithm may become relatively large. The global motion of the system may be characterized by relatively large translational and rotational displacements during a single time step. Therefore, if the numerical integration is not handled properly, the first impact between the colliding bodies is often made with a high penetration depth and, hence, the calculated contact forces become artificially large.

The importance of the initial penetration control, in the framework of the integration of the equations of motion, is better discussed using a simple example. Take the case of the falling ball illustrated in Fig. 3.12, with a mass $m = 1.0\,kg$, a moment of inertia equal to $0.1\,kg\,m^2$, a radius $R = 0.1\,m$, animated by an initial horizontal velocity $v = 1.0\,m/s$ and acted upon by gravity forces only. The motion of the ball is such that during its falling trajectory it strikes the ground. The penetration of the ball in the ground, in the integration time step, for which contact is first detected, is

$$\delta^{(-)} = (h - R) - y_b \tag{3.18}$$

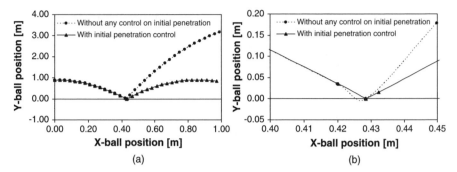

Fig. 3.12 (**a**) Trajectory of a falling ball obtained with integration algorithms with and without initial penetration control; (**b**) detailed view in the vicinity of contact

where y_b is the y coordinate of the ball's center of mass. The superscript $(-)$ on δ means that it is the penetration when contact is first detected. Note that $\delta^{(-)}$ must be a positive value for contact. Therefore, by monitoring the sign of the penetration at every time step $t + \Delta t$ the start can be identified from

$$\delta^{(-)} (\mathbf{q}, t) \, \delta^{(-)} (\mathbf{q}, t + \Delta t) \leq 0 \tag{3.19}$$

When (3.19) is verified the start of contact is defined as occurring at $t + \Delta t$. The integration of the equations of motion of the system could be proceeded with no numerical problem if the penetration first detected is close to zero, or at least below a pre-defined threshold, i.e., if $\delta^{(-)} (\mathbf{q}, t + \Delta t) \leq \delta_{max}$. Because this is not always the case, strategies to limit the time step in the vicinity of contact must be implemented when solving contact problems.

Define as δ^- the distance between the two surfaces in the time step t^- that precedes the time step t^+, at which penetration δ^+ is first detected. In between these time steps, say at t^c, the penetration $\delta^c = 0$ exists. Assuming constant velocity for the multibody system in the vicinity of contact, the time at which contact starts can be calculated by

$$t^c = t^- + \frac{\delta^-}{\delta^+ - \delta^-} \Delta t \tag{3.20}$$

Consequently the ideal situation, during the integration of the multibody equations of motion, would be a time step in the vicinity of contact of

$$\Delta t^{ideal} = t^c - t^- + \varepsilon \tag{3.21}$$

where ε is a very small number to effectively ensure that $\delta_{max} > \delta^c > 0$. Several procedures are suggested to ensure that $\delta^+ < \delta_{max}$, which can be implemented in any code, depending on the access that exists to the numerical integrator.

Procedure 1: Assume that in the vicinity of contact the motion of the multibody system is such that each body moves approximately with constant velocity. Note that the assumption only needs to be valid within a simple time step. Then the time for contact is calculated using (3.20) and the ideal time step using (3.21). Now the positions and velocities of the multibody system, at the time of contact t^c, are estimated as

$$\mathbf{q}^c = \mathbf{q}^- + (\mathbf{q}^+ - \mathbf{q})\frac{\Delta t^{ideal}}{\Delta t} \tag{3.22}$$

$$\dot{\mathbf{q}}^c = \dot{\mathbf{q}}^- + (\dot{\mathbf{q}}^+ - \dot{\mathbf{q}})\frac{\Delta t^{ideal}}{\Delta t} \tag{3.23}$$

where the superscripts $-$, $+$ and c mean that the quantity in which they are applied is evaluated at the instant before contact, after contact and at the time of contact, respectively. The integration algorithm is now restarted at time t^c with the initial positions and velocities given by (3.22) and (3.23).

The procedure proposed, being approximate, does present slight violations of the position and velocity constraint equations. Because a constraint stabilization method or a constraint elimination method is being used, according to the discussion in Chap. 2, it is expected that such violations remain under control. Notice also that when a variable time-step integrator is used during its start the time steps are small. Therefore in the vicinity of contact, small time steps are used by the integrator and even if the conditions calculated by (3.22) and (3.23) are just before contact the integration process continues with the guarantee that the initial penetration never exceeds the prescribed threshold.

Procedure 2: The numerical algorithms used for integration of first-order differential equations with variable time steps, such as the ones generally used in multibody dynamics (Shampine and Gordon 1975, Gear 1981), include an error control that supports the acceptance of rejection of a particular time step. Such decision is based on numerical issues, related to the dynamic response of the system, rather than on any physical reason. The idea behind this procedure, to handle the control on the initial penetration, is to devise a complementary control for the selection of the integration time step based on physical reasoning only. Say that at a given time, during the integration of the equations of motion of multibody system, the internal numerical control of the integration algorithm tests a time step Δt_{trial} and decides to accept it. Before it is definitely accepted, the following physical condition must be met by all new contacts detected in the system:

$$\delta^{(-)}\left(\mathbf{q}, t + \Delta t_{trial}\right) < \delta_{max} \tag{3.24}$$

If the condition described by (3.24) is met by all new contacts, the integration continues without any further interference. If (3.24) is not met the integration algorithm takes it as an indication to reject the time step and attempts a smaller time step. Generally such action corresponds to halving the attempted time step, but particular integration error controls may take different actions. When a smaller new time step is attempted the condition defined by (3.24) is checked again and a decision is made. Eventually a suitable time step that ensures the fulfillment of (3.24) for all new contacts is identified. The integrators available in math libraries include features to inform to the user if the error control intends to accept or reject a time step before doing it. When such features are available the procedure just described is easily implemented.

3.6 Summary

In this chapter, different continuous contact-impact force models for both spherical and cylindrical shape surface collisions in multibody mechanical systems were reviewed. In addition, various types of friction force models based on the Coulomb's law were also listed and discussed. Because modeling contact forces plays a crucial role in the analysis of multibody mechanical systems that experience impacts, the contact force model must be computed using suitable constitutive laws that take into account material properties of the colliding bodies, geometric characteristics of the impacting surfaces and, eventually, the impact velocity. Additionally the numerical method for the calculation of the contact force should be stable enough to allow for the integration of the mechanical equations of motion with acceptable efficiency. These characteristics are ensured by using a continuous contact force model in which the force and penetration vary in a continuous manner and for which some energy dissipation is included. This approach has the extra benefit of leading to a behavior of the variable time-step integrator that is more stable.

Some important conclusions can be drawn from the study presented in this chapter. Among the spherical-shaped contact areas, the linear Kelvin–Voigt contact model does not represent the overall nonlinear nature of impact phenomenon. The Hertz relation does not account for the energy dissipation during the impact process. Therefore the Hertz relation, along with the modification to represent the energy dissipation, in the form of internal damping, can be adopted for modeling contact forces in a multibody system. This model is straightforward and easy to implement in a computational program.

The cylindrical models are nonlinear and implicit functions, and therefore, they require a numerical iterative procedure to be performed. Furthermore these models have been proposed as purely elastic, not being able to explain the energy dissipation during the impact process. From the comparison between the spherical and cylindrical contact force models, it can be concluded that the spherical and cylindrical force–displacement relations are reasonably close. Therefore, the straightforward force–penetration relation proposed by Lankarani and Nikravesh (1990), with a modification of the pseudo-stiffness parameter in the case of cylindrical contact, is largely used for mechanical contacts owing to its simplicity and easiness of implementation in a computational program. Aiding to those advantages, this is the only model that accounts for energy dissipation during the impact process.

In dynamic analysis of multibody systems, the deformation/indentation is known at every time step from the configuration of the system, the forces evaluated being based on the state variables. With the variation of the contact force during the contact period, the dynamic system response is obtained by simply including updated forces into the equations of motion. Since the equations of motion are integrated over the period of contact, this approach results in a rather accurate response. This procedure was further improved by including in the time integration scheme a procedure that controls the time step in order to prevent large penetrations to develop in the initial contact. Furthermore this methodology accounts for the changes in the system's configuration during the contact period. This approach is employed in the forthcoming chapters to describe the impact between the elements that compose the joint clearances.

References

Ahmed S, Lankarani HM, Pereira MFOS (1999) Frictional impact analysis in open loop multibody mechanical system. Journal of Mechanical Design 121:119–127.

Ambrósio J (2000) Rigid and flexible multibody dynamics tools for the simulation of systems subjected to contact and impact conditions. European Journal of Solids A/Solids 19:S23–S44.

Ambrósio J (2002) Impact of rigid and flexible multibody systems: deformation description and contact models. Virtual nonlinear multibody systems, NATO Advanced Study Institute, Prague, Czech Republic, June 23–July 3, edited by W Schiehlen and M Valásek, Vol. II, pp. 15–33.

Bagci C (1975) Dynamic motion analysis of plane mechanisms with Coulomb and viscous damping via the joint force analysis. Journal of Engineering for Industry, Series B 97(2):551–560.

Baumgarte J (1972) Stabilization of constraints and integrals of motion in dynamical systems. Computer Methods in Applied Mechanics and Engineering 1:1–16.

Bottasso CL, Citelli P, Taldo A, Franchi CG (1999) Unilateral contact modeling with adams. International ADAMS user's conference, Berlin, Germany, November 17–18, 11pp.

Brach RM (1991) Mechanical impact dynamics, rigid body collisions. Wiley, New York.

Dietl P (1997) Damping and stiffness characteristics of rolling element bearings—theory and experiment. Ph.D. Dissertation, Technical University of Vienna, Austria.

Dubowsky S (1974) On predicting the dynamic effects of clearances in planar mechanisms. Journal of Engineering for Industry, Series B 96(1):317–323.

Dubowsky S, Freudenstein F (1971) Dynamic analysis of mechanical systems with clearances, part 1: formulation of dynamic model. Journal of Engineering for Industry, Series B 93(1):305–309.

ESDU 78035 Tribology Series (1978) Contact phenomena. I: stresses, deflections and contact dimensions for normally loaded unlubricated elastic components. Engineering Sciences Data Unit, London, England.

Flores P, Ambrósio J, Claro JCP, Lankarani HM (2006) Influence of the contact-impact force model on the dynamic response of multibody systems. Proceedings of the Institution of Mechanical Engineers, Part-K Journal of Multi-body Dynamics 220:21–34.

Gear CW (1981) Numerical solution of differential-algebraic equations. IEEE Transactions on Circuit Theory CT-18:89–95.

Goldsmith W (1960) Impact—the theory and physical behaviour of colliding solids. Edward Arnold, London, England.

Greenwood DT (1965) Principles of dynamics. Prentice Hall, Englewood Cliffs, NJ.

Han I, Gilmore BJ (1993) Multi body impact motion with friction analysis, simulation, and validation. Journal of Mechanical Design 115:412–422.

Haug EJ, Wu SC, Yang SM (1986) Dynamics of mechanical systems with Coulomb friction, stiction, impact and constraint addition deletion—I theory. Mechanism and Machine Theory 21(5):401–406.

Hertz H (1896) On the contact of solids—on the contact of rigid elastic solids and on hardness. Miscellaneous papers (Translated by DE Jones and GA Schott), pp. 146–183. Macmillan, London, England.

Hunt KH, Crossley FR (1975) Coefficient of restitution interpreted as damping in vibroimpact. Journal of Applied Mechanics 7:440–445.

Keller JB (1986) Impact with friction. Journal of Applied Mechanics 53:1–4.

Khulief YA, Shabana AA (1986) Dynamic analysis of constrained system of rigid and flexible bodies with intermittent motion. Journal of Mechanisms, Transmissions, and Automation in Design 108:38–45.

Lankarani HM (1988) Canonical equations of motion and estimation of parameters in the analysis of impact problems. Ph.D. Dissertation, University of Arizona, Tucson, AZ.

Lankarani HM (2000) A poisson based formulation for frictional impact analysis of multibody mechanical systems with open or closed kinematic chains. Journal of Mechanical Design 115:489–497.

Lankarani HM, Nikravesh PE (1990) A contact force model with hysteresis damping for impact analysis of multibody systems. Journal of Mechanical Design 112:369–376.

Lankarani HM, Nikravesh PE (1994) Continuous contact force models for impact analysis in multibody systems. Nonlinear Dynamics 5:193–207.

Love AEH (1944) A treatise on the mathematical theory of elasticity, Fourth edition. Dover Publications, New York.

Maw N, Barber JR, Fawcett JN (1976) The oblique impact of elastic spheres. Wear 38:101–114.

Nikravesh PE (1988) Computer-aided analysis of mechanical systems. Prentice Hall, Englewood Cliffs, NJ.

PC-Crash (2002) A simulation program for vehicle accidents, Technical manual, Version 6.2.

Pedersen S, Hansen J, Ambrósio J (2002) A novel roller-chain drive model using multibody dynamics analysis tools. Virtual nonlinear multibody systems, NATO Advanced Study Institute, Prague, Czech Republic, June 23–July 3, edited by W Schiehlen and M. Valásek, Vol. II, pp. 180–185.

Pedersen SL (2001) Chain vibrations. MS Dissertation, Department of Mechanical Engineering, Solid Mechanics, Technical University of Denmark, Lyngby, Denmark.

Pfeiffer F, Glocker C (1996) Multibody dynamics with unilateral constraints. Wiley, New York.

Ravn P (1998) A continuous analysis method for planar multibody systems with joint clearance. Multibody System Dynamics 2:1–24.

Rooney GT, Deravi P (1982) Coulomb friction in mechanism sliding joints. Mechanism and Machine Theory 17:207–211.

Ryan RR (1990) ADAMS—multibody system analysis software. Multibody systems handbook. Springer, Berlin Heidelberg New York.

Shampine L, Gordon M (1975) Computer solution of ordinary differential equations: the initial value problem. Freeman, San Francisco, CA.

Shivaswamy S (1997) Modeling contact forces and energy dissipation during impact in multibody mechanical systems. Ph.D. Dissertation, Wichita State University, Wichita, KS.

Silva MPT, Ambrósio J (2004) Human motion analysis using multibody dynamics and optimization tools. Technical report IDMEC/CPM—2004/001, Instituto Superior Técnico of the Technical University of Lisbon, Lisbon, Portugal.

Smith RC, Haug EJ (1990) DADS—dynamic analysis and design system. Multibody systems handbook. Springer, Berlin Heidelberg New York.

Stolarski TA (1990) Tribology in machine design. Butterworth-Heinemann, Oxford, England.

Threlfall DC (1978) The inclusion of Coulomb friction in mechanisms programs with particular reference to DRAM. Mechanism and Machine Theory 13:475–483.

Timoshenko SP, Goodier JN (1970) Theory of elasticity. McGraw-Hill, New York.

Wang Y, Manson M (1992) Two dimensional rigid-body collisions with friction. Journal of Applied Mechanics 59:635–642.

Wilson R, Fawcett JN (1974) Dynamics of slider–crank mechanism with clearance in the sliding bearing. Mechanism and Machine Theory 9:61–80.

Wu SC, Yang SM, Haug EJ (1986a) Dynamics of mechanical systems with Coulomb friction, stiction, impact and constraint addition deletion—II planar systems. Mechanism and Machine Theory 21(5):407–416.

Wu SC, Yang SM, Haug EJ (1986b) Dynamics of mechanical systems with Coulomb friction, stiction, impact and constraint addition deletion—III spatial systems. Mechanism and Machine Theory 21(5):417–425.

Zhu SH, Zwiebel S, Bernhardt G (1999) A theoretical formula for calculating damping in the impact of two bodies in a multibody system. Proceedings of the Institution of Mechanical Engineers 213(C3):211–216.

Zukas JA, Nicholas T, Greszczuk LB, Curran DR (1982) Impact dynamics. Wiley, New York.

Chapter 4
Planar Joints with Clearance: Dry Contact Models

In general, a multibody system is made of several components, which can be divided into links with a convenient geometry and joints, which introduce restrictions on the relative motion of the various bodies of the system (Shabana 1989). Usually the links are modeled as rigid or deformable bodies, while the joints are modeled through a set of kinematic constraints, that is, the joints are not modeled *de per si* (Bauchau and Rodriguez 2002). The functionality of a kinematic joint relies upon the relative motion allowed between the connected components, which, in practice, implies the existence of a gap, that is, a clearance between the mating parts, leading to surface contact, shock transmission and the development of different regimes of friction and wear. No matter how small the clearance is, it can lead to vibration and fatigue phenomena, lack of precision or, in the limit, to even random overall behavior. If there is no lubricant or other damping materials in the joint, impacts occur in the system and the corresponding impulses are transmitted throughout the multibody system. In this work, the elements that compose the clearance joints are modeled as colliding bodies. The impact between the two bodies is treated as a continuous event, that is, the local deformations and the contact forces are continuous functions of time. The impact analysis of the system is performed simply by including the contact-impact forces in the equations of motion during the impact period (Flores and Ambrósio 2004). From the system configuration, a geometric condition defines if the elements of the joints are in contact or not. Thus the dynamics of joints with clearances is controlled by contact-impact forces, rather than by the kinematic constraints of the ideal joints. A force model that accounts for the geometric and material characteristics of the clearance joint components describes these impacts and the eventual continuous contact (Lankarani and Nikravesh 1990). The energy dissipative effects are introduced in the joints through the contact force model and by friction forces that develop during the contact (Ambrósio 2002). The main purpose of this chapter is to present the mathematical models for revolute and translational joints with clearance in planar multibody mechanical systems. The methodologies and procedures adopted in this chapter are applied to a slider–crank mechanism, which includes both revolute and translational joints with clearance. For the case of revolute joints with clearance, the contact force models presented in the previous chapter are used to assess the influence of the different models on the impact force.

P. Flores et al., *Kinematics and Dynamics of Multibody Systems with Imperfect Joints.*   67
© Springer 2008

4.1 Clearance Models

It is known that the performance of a multibody system is degraded by the presence of clearances in the joints because impact forces occur. These impact forces contribute to the failure of the components due to shock loading, reducing the systems' life due to material fatigue, generating high noise levels, causing energy dissipation and exciting unwanted vibratory responses (Dubowsky and Freudenstein 1971a,b, Ravn 1998, Flores and Ambrósio 2004, Flores et al. 2006).

In standard multibody models, it is assumed that the connecting points of two bodies, linked by an ideal or perfect revolute joint, are coincident. The introduction of the clearance in a revolute joint allows for the separation of these two points. Figure 4.1 depicts a revolute joint with clearance, that is, the so-called journal–bearing, where the difference in radius between the bearing and the journal defines the radial clearance.

Although a revolute joint with clearance does not constrain any degree of freedom from the mechanical system, as the ideal joint does, it imposes some kinematic restrictions, limiting the journal to move within the bearing. Thus, when the clearance is present in a planar revolute joint, two kinematic constraints are removed and two degrees of freedom are introduced instead. The dynamics of the joint is then controlled by contact-impact forces between the journal and the bearing. Thus, whilst a perfect revolute joint in a mechanical system imposes kinematic constraints, a revolute clearance joint leads to force constraints. Therefore mechanical joints with clearance can be defined as force-joints instead of kinematic joints.

In a revolute clearance joint, when contact exists between the journal and the bearing, a contact-impact force, perpendicular to the plane of collision, develops. The force is typically applied as a spring–damper element. If this element is linear, the approach is known as the Kelvin–Voigt model (Timoshenko and Goodier 1970). If the relation is nonlinear, the model is generally based on the Hertz contact law (Hertz 1896).

In what concerns the clearance modeling, there are, in general, three different approaches, namely, the massless link approach, the spring–damper approach and the

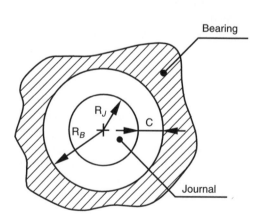

Fig. 4.1 Revolute joint with clearance, that is, the so-called journal–bearing

momentum exchange approach. The massless link approach (Earles and Wu 1975), in which the presence of clearance at a joint is modeled by adding an imaginary massless link with a fixed length equal to the clearance, is illustrated in Fig. 4.2a. This link results in the mechanism model having an additional degree of freedom. Hence the resulting equations of motion are found to be highly nonlinear and complex to solve. Furthermore this model assumes that there is contact between the journal and the bearing all the time, hence unable to represent free flight trajectories. Wu and Earles (1977) used the massless link model to predict the occurrence of contact loss in revolute joints of planar mechanisms.

The spring–damper approach (Dubowsky and Freudenstein 1971a,b, Bengisu et al. 1986), in which the clearance is modeled by introducing a spring–damper element that simulates the surface elasticity, is in Fig. 4.2b. This model shows some deficiencies in representing the physical nature of the energy transfer during the impact process, the parameters of the spring and damper elements being difficult to quantify. Dubowsky (1974) investigated the dynamic effects of clearance in planar mechanisms by simulating the elasticity of the contacting surfaces using linear springs and dampers.

In the momentum exchange approach (Townsend and Mansour 1975, Ravn 1998, Flores et al. 2006), the mechanical elements that constitute a clearance joint are considered as impacting bodies. The contact-impact forces control the dynamics of the clearance joint. The work presented in this book uses methodologies that are in line with momentum exchange approach in the kinematics of the contacting bodies concerned. In the massless link and spring–damper models, the clearance is replaced by mechanical components, which are intended to represent the behavior of the clearance as closely as possible. The momentum exchange approach is more realistic since the impact force model allows, with high level of approximation, to simulate the elasticity of the contacting surfaces as well as the energy dissipation during the impact.

Several published works focused on the different modes of motion of the journal inside the bearing. Most of these consider a three-mode model for predicting the dynamical response of articulated systems with revolute clearance joints (Mansour and Townsend 1975, Miedema and Mansour 1976). The three different modes of journal motion inside the bearing are the contact or following mode, the free flight

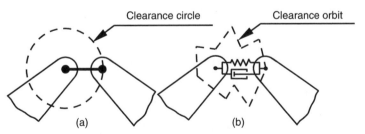

Fig. 4.2 Examples of models for revolute joints with clearance: (**a**) massless link model; (**b**) spring–damper model

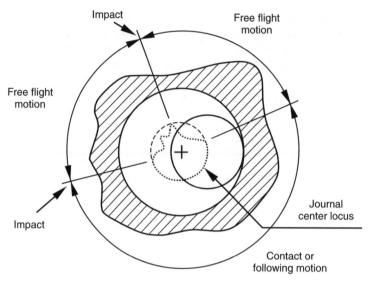

Fig. 4.3 Types of journal motion inside the bearing

mode and the impact mode. These three types of the journal motion are illustrated in Fig. 4.3.

In the contact or following mode, the journal and the bearing are in permanent contact and a sliding motion between the contacting surfaces exists. In this mode, the relative penetration depth varies along the circumference of the bearing. This mode ends when the journal and the bearing separate from each other, and the journal enters the free flight mode. In the free flight mode, the journal can move freely inside the bearing boundaries, that is, the journal and the bearing are not in contact and, consequently, no reaction force develops at the joint. In the impact mode, which occurs on the termination of the free flight mode, impact forces are applied to the system. This mode is characterized by a discontinuity in the kinematic and dynamic characteristics, and a significant exchange of momentum occurs between the two impacting bodies. At the termination of the impact mode, the journal can enter either a free flight or a following mode. During the dynamic simulation of a revolute joint with clearance, if the path of the journal center is plotted for each instant, these different modes of motion, depicted in Fig. 4.3, can be easily identified.

4.2 Model of Revolute Joint with Clearance

The simulation of real joints requires the development of a mathematical model for revolute clearance joints in the multibody systems. Figure 4.4 shows two bodies i and j connected by a generic revolute joint with clearance. Part of body i is the bearing and part of body j is the journal. The center of mass of bodies i and j are O_i and O_j, respectively. Body-fixed coordinate systems $\xi\eta$ are attached to the

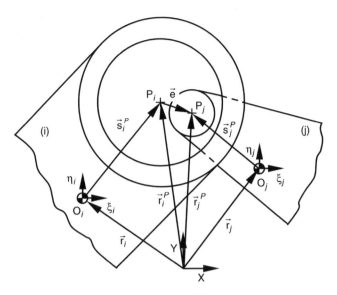

Fig. 4.4 Generic revolute joint with clearance in a multibody system

center of mass of each body, while the XY coordinate frame represents the global coordinate system. Point P_i indicates the center of the bearing, and the center of the journal is denominated by point P_j.

In the dynamic simulation, the behavior of the revolute clearance joint is treated as an oblique eccentric impact between the journal and the bearing. The mechanics of this type of impact involves both the relative normal velocity and the relative tangential velocity (Zukas et al. 1982). When the impact occurs, an appropriate contact law must be applied, the resulting forces being introduced in the system equations of motion as generalized forces.

Taking into account Fig. 4.4, the eccentricity vector **e** connecting the centers of the bearing and the journal is calculated as

$$\mathbf{e} = \mathbf{r}_j^P - \mathbf{r}_i^P \tag{4.1}$$

where both $\mathbf{r}_i^P$ and $\mathbf{r}_j^P$ are described in the global coordinates reference frame as (Nikravesh 1988)

$$\mathbf{r}_k^P = \mathbf{r}_k + \mathbf{A}_k \mathbf{s}_k^{'P}, \, (k = i, j) \tag{4.2}$$

The magnitude of the eccentricity vector is evaluated as

$$e = \sqrt{\mathbf{e}^T \mathbf{e}} \tag{4.3}$$

where $\mathbf{e}^T$ is the transpose of vector **e**.

The unit normal vector **n** to the surfaces in collision between the bearing and the journal, designated as **n**, is aligned with the eccentricity vector:

$$\mathbf{n} = \mathbf{e}/e \tag{4.4}$$

The unit vector is aligned with the line between the centers of the bearing and the journal.

With reference to Fig. 4.5, the penetration depth caused by the impact between the journal and the bearing is evaluated as

$$\delta = e - c \tag{4.5}$$

where c is the radial clearance, defined as the difference between the radius of the bearing and the radius of the journal.

Let points Q_i and Q_j represent the contact points on bodies i and j, respectively. The position of the contact points Q_i and Q_j are evaluated as

$$\mathbf{r}_k^Q = \mathbf{r}_k + \mathbf{A}_k \mathbf{s}_k'^Q + R_k \mathbf{n}, \ (k = i, j) \tag{4.6}$$

where R_i and R_j are the bearing and journal radiis, respectively.

In some contact models, it is important to evaluate the dissipative effects that develop during impact. In the continuous force contact model it, is necessary to calculate the relative velocity between the impacting surfaces. The velocity of the contact points Q_i and Q_j in the global coordinate system is found by differentiating (4.6) with respect to time, i.e.,

$$\dot{\mathbf{r}}_k^Q = \dot{\mathbf{r}}_k + \dot{\mathbf{A}}_k \mathbf{s}_k'^Q + R_k \dot{\mathbf{n}} \tag{4.7}$$

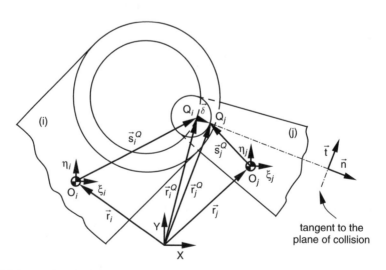

Fig. 4.5 Penetration depth due to the impact between the bearing and the journal

where ($\bullet$) denotes the derivative with respect to time of quantity ($\bullet$).

The relative velocity between the contact points is projected onto the tangential line to the colliding surfaces and onto the normal to colliding surfaces, yielding a relative tangential velocity, $\mathbf{v}_T$, and a relative normal velocity, $\mathbf{v}_N$, shown in Fig. 4.6. The normal relative velocity determines whether the contact bodies are approaching or separating. The tangential relative velocity determines whether the contact bodies are sliding or sticking. The relative scalar normal and tangential velocities are

$$v_N = (\dot{\mathbf{r}}_j^Q - \dot{\mathbf{r}}_i^Q)^T \mathbf{n} \qquad (4.8)$$

$$v_T = (\dot{\mathbf{r}}_j^Q - \dot{\mathbf{r}}_i^Q)^T \mathbf{t} \qquad (4.9)$$

where $\mathbf{t}$ is obtained by rotating the vector $\mathbf{n}$, calculated using (4.4), in the counter clockwise direction by $90°$.

The normal and tangential forces, $\mathbf{f}_N$ and $\mathbf{f}_T$, respectively, are applied at the contact points. These forces are evaluated using the contact force law proposed in Chap. 3 and a friction model such as the Coulomb law. The contributions to the generalized vector of forces and moments, $\mathbf{g}$ in the equation of motion, are found by projecting the normal and tangential forces onto the X and Y directions. These forces that act on the contact points of bodies i and j are transferred to the center of mass of bodies and an equivalent transport moment is applied to the rigid body. Referring to Fig. 4.7, the forces and moments that act on the center of mass of body i due to the clearance joint contact are given by

$$\mathbf{f}_i = \mathbf{f}_N + \mathbf{f}_T \qquad (4.10)$$

$$m_i = -(y_i^Q - y_i)f_i^x + (x_i^Q - x_i)f_i^y \qquad (4.11)$$

The corresponding forces and moments applied to the body j are

$$\mathbf{f}_j = -\mathbf{f}_i \qquad (4.12)$$

$$m_j = (x_j^Q - x_j)f_j^y - (y_j^Q - y_j)f_j^x \qquad (4.13)$$

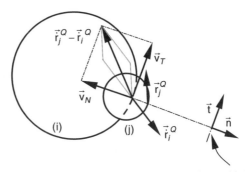

Fig. 4.6 Velocity vectors of impact between the bearing and the journal

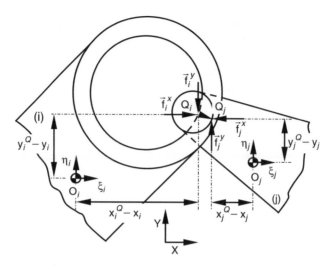

Fig. 4.7 Force vectors that act at the points of contact

4.3 Model of Translational Joint with Clearance

Similar to the procedure used in the previous section for the revolute joint, this presents a mathematical model for translational joints with clearance in multibody systems. Figure 4.8 shows an example of a planar translational joint with clearance. The clearance c is defined as the difference between the guide and slider surfaces. The geometric characteristics of the translational clearance joint used here are the length of the slider L, the slider width W and the distance between the guide surfaces H. In the present work, the slider and the guide elements that constitute a translational clearance joint are modeled as two colliding bodies and the dynamics of the joint is governed by contact-impact forces. The equations of motion that govern the dynamic response of the general multibody systems incorporate these forces.

   In an ideal translational joint the two bodies, slider and guide, translate with respect to each other along the line of translation, so that there is neither rotation between the bodies nor a relative translation motion in the direction perpendicular to the axis of the joint. Therefore an ideal translational joint reduces the number

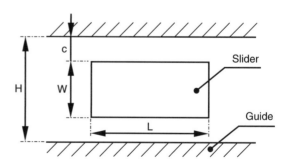

Fig. 4.8 Translational joint with clearance, that is, the slider and its guide

of degrees of freedom of the system by two. The existence of a clearance in a translational joint removes the two kinematic constraints and introduces two extra degrees of freedom. Hence the slider can move 'freely' inside the guide limits. When the slider reaches the guide surfaces an impact occurs and the dynamic response of the joint is controlled by contact forces. These contact forces are evaluated according to the continuous contact force proposed together with the dissipative friction force model selected. Then these forces are introduced into the system equations of motion as external generalized forces. Although a translational clearance joint does not constrain any degree of freedom from the mechanical system, as an ideal joint does, it imposes some restrictions on the slider motion inside the guide. Thus, while a perfect joint in a multibody system is achieved by kinematic constraints, a clearance joint is obtained by force constraints.

Over the last few decades extensive work has been done to study the dynamic effect of the revolute clearance in multibody systems (Dubowsky and Freudenstein 1971a, Ravn 1998, Schwab 2002, Flores and Ambrósio 2004). In contrast, little work has been done to model translational joints with clearance. Wilson and Fawcett (1974) derived the equations of motion for the different scenarios of the slider motion inside the guide. They also showed how the slider motion in a translational clearance joint depends on the geometry, speed and mass distribution. Farahanchi and Shaw (1994) studied the dynamic response of a planar slider–crank mechanism with slider clearance. More recently, Thümmel and Funk (1999) used the complementary approach to model impact and friction in a slider–crank mechanism with both revolute and translational clearance joints.

The modeling of translational clearance joints is more complicated than that for the revolute joints, due to the several possible contact configurations between the slider and the guide. Figure 4.9 illustrates four different scenarios for the slider configuration relative to guide surface, namely: (1) no contact between the two elements: the slider is in free flight motion inside the guide and, consequently, there is no reaction force at the joint; (2) one corner of the slider is in contact with the guide surface; (3) two adjacent slider corners are in contact with the guide surface, which corresponds to having a face of slider in contact with the guide surface; (4) two opposite slider corners are in contact with the guide surface. The conditions for switching between different cases depend on the system's dynamics. For the cases represented in Fig. 4.9, the contact forces are evaluated using the continuous contact force model.

In order for the clearance joints to be simulated in the multibody system environment, it is required that a mathematical model be developed. Figure 4.10 shows a representation of a translation joint with clearance that connects bodies i and j. The slider is body i whereas the guide is part of body j. The center of mass of bodies i and j are O_i and O_j, respectively. Let points P, Q, R and S in the guide surfaces indicate the geometric limits inside which contact may occur. Points A_i, B_i, C_i and D_i indicate the four corners of the slider, and A_j, B_j, C_j and D_j are the points on the guide surfaces that are closer to the points in body j. The contact formulation for all corners in the slider is similar, and therefore, in what follows only the slider corner A is used to describe the mathematical formulation.

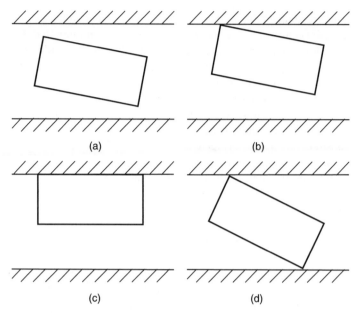

Fig. 4.9 Different scenarios for the slider motion inside the guide: (**a**) no contact; (**b**) one corner in contact with the guide; (**c**) two adjacent corners in contact with the guide; (**d**) two opposite corners in contact with the guide

Let vector $\mathbf{t}$, directed along the guide surface from point P to point Q in body j, be written in terms of the body-fixed coordinates as

$$\mathbf{t}'_j = \mathbf{s}'^Q_j - \mathbf{s}'^P_j \tag{4.14}$$

Note that the tangent vector expressed in the inertia frame is $\mathbf{t} = \mathbf{A}_j \mathbf{t}'$, where $\mathbf{A}_j$ is the transformation matrix from body j's frame to the inertial frame.

Let the position vector for any given point G of a body k be described with respect to inertial reference frame as

$$\mathbf{r}^G_k = \mathbf{r}_k + \mathbf{A}_k \mathbf{s}'^G_k, \ (k = i, j) \tag{4.15}$$

where $\mathbf{s}'^G_k$ is the position of point G in body k expressed in body-fixed coordinates.

The position of point A_j, belonging to the segment PQ of the guide, closest to point A_i located in the corner of the slider, is given as

$$\mathbf{r}^A_j = \mathbf{r}^P_j + [\mathbf{t}^T (\mathbf{r}^A_i - \mathbf{r}^P_j)]\mathbf{t} \tag{4.16}$$

The vector connecting the slider corner A_i to point A_j on the guide surface is

$$\delta = \mathbf{r}^A_j - \mathbf{r}^A_i \tag{4.17}$$

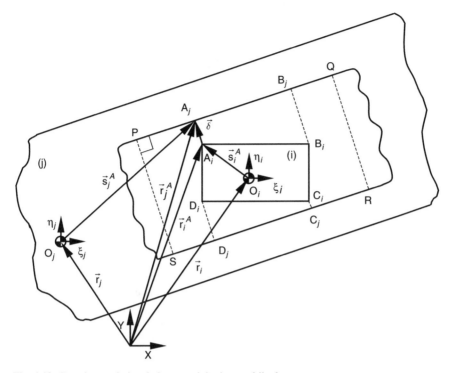

Fig. 4.10 Generic translational clearance joint in a multibody system

Note that vector $\boldsymbol{\delta}$ has the same direction as the normal **n** to the guide surface. Regardless of this identity, let the normal vector **n** be defined as perpendicular to the tangent vector **t**, which for two-dimensional cases is

$$\mathbf{n} = \begin{bmatrix} t_y & -t_x \end{bmatrix}^T \tag{4.18}$$

Figure 4.11 shows the slider and guide in two different scenarios, namely in a noncontact situation and in the case of penetration between the slider and guide surface. For the contact case, the vectors $\boldsymbol{\delta}$ and **n** are parallel but oriented in opposite directions. Thus the condition for penetration between the slider and guide is expressed as

$$\mathbf{n}^T \boldsymbol{\delta} < 0 \tag{4.19}$$

The magnitude of the penetration depth for point A_i is evaluated as

$$\delta = \sqrt{\boldsymbol{\delta}^T \boldsymbol{\delta}} \tag{4.20}$$

The impact velocity, required for the evaluation of the contact force, is obtained by differentiating (4.17) with respect to time, yielding

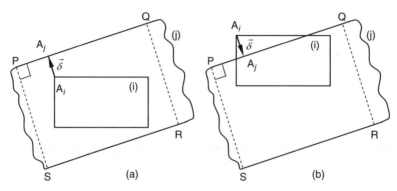

Fig. 4.11 (a) Noncontact situation; (b) penetration between the slider corner A and the guide surface

$$\dot{\boldsymbol{\delta}} = \dot{\mathbf{r}}_j + \dot{\mathbf{A}}_j \mathbf{s}'^A_j - \dot{\mathbf{r}}_i - \dot{\mathbf{A}}_i \mathbf{s}'^A_i \tag{4.21}$$

When the contact between the slider and the guide surfaces takes place, normal and tangential forces appear in the contact points. By transferring these forces to the center of mass of each body Fig. 4.12 shows the forces and moments acting on the center of mass of body I, which are

$$\mathbf{f}_i = \mathbf{f}_N + \mathbf{f}_T \tag{4.22}$$

$$m_i = -(y^Q_i - y_i) f^x_i + (x^Q_i - x_i) f^y_i \tag{4.23}$$

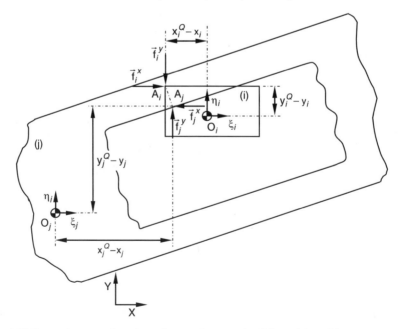

Fig. 4.12 Forces that act at the points of contact between the slider and the guide

The forces and moments to be applied in body j are written as

$$\mathbf{f}_j = -\mathbf{f}_i \tag{4.24}$$
$$m_j = (x_j^Q - x_j)f_j^y - (y_j^Q - y_j)f_j^x \tag{4.25}$$

In dealing with translational clearance joints, it is essential to define how the slider and guide surfaces contact each other and, consequently, what is the most adequate contact force model. Lankarani (1988) presented a linear model for contact between two square plane surfaces as

$$F_N = K\delta \tag{4.26}$$

where the stiffness parameter K is given by

$$K = \frac{a}{0.475(\sigma_i + \sigma_j)} \tag{4.27}$$

having the area of contact a length of $2a$ and quantities σ_i and σ_j are given by (3.4).

When two adjacent slider corners contact with the guide surface, the resulting contact force is applied at the geometric center of the penetration area, denoted as GC in Fig. 4.13a, and the contact force model given by (4.26) is used. Otherwise, when one or two opposite slider corners contact the guide surface the contact is assumed to be between a spherical surface and a plane surface, allowing for the contact model given by Hertz law with hysteretic damping factor expressed by (3.9) to be applied. In order to evaluate the equivalent stiffness, a small curvature radius R_c is assumed on the contact corner, represented in Fig. 4.13b. The unified contact model is obtained using the pseudo-stiffness expressed by (4.27) in the continuous force model represented by (3.9).

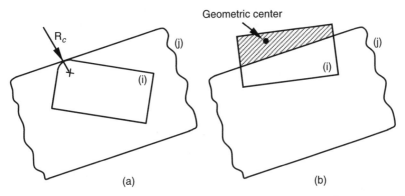

Fig. 4.13 (a) Contact between a spherical surface and a plane; **(b)** contact between two plane surfaces

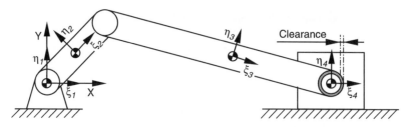

Fig. 4.14 Slider–crank mechanism with a revolute clearance joint

4.4 Application 1: Slider–Crank with Revolute Clearance Joint

The slider–crank mechanism is chosen as an example to demonstrate the application of the methodologies presented in this chapter. The same mechanism has been studied by other authors (Ravn 1998, Schwab 2002), which allows the comparison of the results obtained. Figure 4.14 shows the configuration of the slider–crank mechanism, which consists of four rigid bodies, two ideal revolute joints and one ideal translational joint. A revolute clearance joint exists between the connecting rod and the slider.

The crank, which is the driving link, rotates with a constant angular velocity of 5000 rpm. The initial configuration of the mechanism is defined with the crank and the connecting rod collinear and the journal and bearing centers coincident. Furthermore the initial positions and velocities necessary to start the dynamic analysis are obtained from kinematic simulation of the slider–crank mechanism in which all the joints are considered to be ideal. The geometric and inertia properties of each body are listed in Table 4.1. The parameters used for the different models, required to characterize the problem, and for the numerical methods, required to solve the system dynamics, are listed in Table 4.2.

The dynamic response of the slider–crank mechanism is obtained and represented in Figs. 4.15 and 4.16 by the time plots of the velocity and acceleration of the slider and the moment acting on the crank, which is required to maintain the crank angular velocity constant. The relative motion between journal and bearing centers is plotted in Figs. 4.15d and 4.17. The Hertz contact force law with hysteretic damping factor, given by (3.9), is used to evaluate the contact force between the journal and bearing. Figure 4.15 shows the results for the case in which the clearance size is 0.5 mm. Note that the results, reported for the two full crank rotations after steady-state has been reached, are plotted against those obtained for an ideal joint.

In Fig. 4.15a, it is observed that the existence of a joint clearance influences the slider velocity by leading to a staircase-like behavior. The periods of constant

Table 4.1 Geometric and inertia properties of the slider–crank mechanism

Body nr	Length (m)	Mass (kg)	Moment of inertia (kg m^2)
2	0.05	0.30	0.00010
3	0.12	0.21	0.00025
4	–	0.14	0.00010

Table 4.2 Parameters used in the dynamic simulation of the slider–crank mechanism with revolute clearance joint

Bearing radius	10.0 mm	Baumgarte - α	5
Restitution coefficient	0.9	Baumgarte - β	5
Young's modulus	207 GPa	Integration step	10^{-5} s
Poisson's ratio	0.3	Integration tolerance	10^{-6} s

velocity observed for the slider mean that the journal moves freely inside the bearing boundaries. The sudden changes in velocity are due to the impact between the journal and the bearing. When a smooth change in the velocity curve of the slider is observed it indicates that the journal and the bearing are in continuous contact, that is, the journal follows the bearing wall. This situation is confirmed by smooth changes in the acceleration curve. The slider acceleration exhibits high peaks caused by impact forces that are propagated through the rigid bodies of the mechanism, as perceived in Fig. 4.15b where the acceleration of the slider is displayed. The slider acceleration presents high peaks of its values, which may be smoothed in a real mechanism due to the energy dissipation associated with the system components' flexibility. The same phenomena can be observed in the crank moment, represented by Fig. 4.15c. As far as the trajectory of the journal center relative to the bearing center is concerned, different types of motion between the two bodies can be

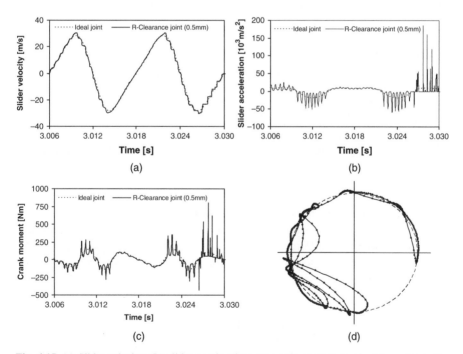

Fig. 4.15 (**a**) Slider velocity; (**b**) slider acceleration; (**c**) crank moment required to maintain its angular velocity constant; (**d**) journal trajectory relative to the bearing

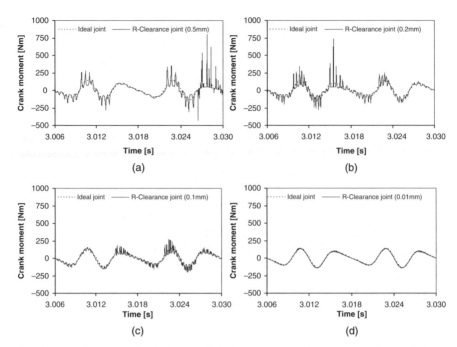

Fig. 4.16 Crank moment for different radial clearance sizes: (**a**) $c = 0.5$ mm; (**b**) $c = 0.2$ mm; (**c**) $c = 0.1$ mm; (**d**) $c = 0.01$ mm

observed, namely, free flight, impact and rebound, and permanent or continuous contact. The relative penetration depth between the journal and the bearing is visible by the points of the journal that are plotted outside the clearance circle, in Fig. 4.15d, where the journal center trajectories are presented by continuous lines that connect point markers. Each one of the markers represents the position of the journal for a given time step. It can be observed that during the free flight motion the time step adopted by the integration algorithm is much larger than during the contact. When contact is detected, the integration time step decreases significantly, which shows the importance of varying time-step integration algorithm for problems involving contact.

The clearance size is one of the most important parameters that affect the dynamic behavior of the system. In Figs. 4.16–4.18, the crank moments, the journal center trajectories and the Poincaré maps are used to illustrate the dynamic behavior of the slider–crank mechanism when different clearance sizes are present. In this application, the slider acceleration and slider velocity are chosen to plot the Poincaré maps. The values for the clearance of the revolute joint are chosen to be 0.5, 0.2, 0.1 and 0.01 mm.

Poincaré maps are mathematical abstractions which are often useful in highlighting the dynamic behavior of systems in terms of periodic, quasi-periodic and chaotic motion. Especially chaotic systems are often examined through the use of Poincaré maps. A Poincaré map consists of plotting the value of two components

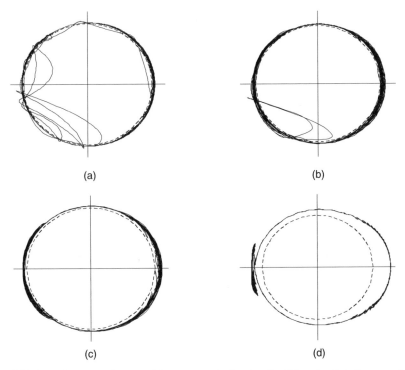

(a)   (b)

(c)   (d)

Fig. 4.17 Journal center trajectory with respect to the bearing for different radial clearance sizes, the maximum eccentricity being plotted with a *dashed circle*: (**a**) $c = 0.5$ mm; (**b**) $c = 0.2$ mm; (**c**) $c = 0.1$ mm; (**d**) $c = 0.01$ mm

from the state vector versus its derivative, i.e., $y(t)$ and $\dot{y}(t)$ (Baker and Gollub 1990, Tomsen 1997). Regular or periodic behavior is insensitive to initial conditions and is represented in the Poincaré map by a closed orbit or finite number of points. Chaotic or nonperiodic responses are extremely sensitive to initial conditions and are perceived by a region densely filled by orbits or points in the Poincaré map. A complicated looking phase in a Poincaré map is one indicator of chaotic motion. Quasi-periodic orbits fill up the Poincaré maps as the chaotic orbits, but they do so in a fully predictable manner since there is not such a sensitive dependency on the initial conditions (Wiggins 1990).

From the Poincaré maps analysis, the slider–crank mechanism behavior can easily be characterized, and it is possible to distinguish between periodic, quasi-periodic and chaotic responses. In multibody systems, nonlinearities arise from intermittent motion, clearance joints, friction effect and contact forces, among others. The relation between the clearance size and the type of motion observed is clearly identified from plots in Fig. 4.18.

Figure 4.16 shows that when the clearance size is decreased the dynamic behavior tends to be smoother, which is represented by lower peaks in the crank moment. Indeed, when the clearance is small, the system response tends to be closer to the ideal

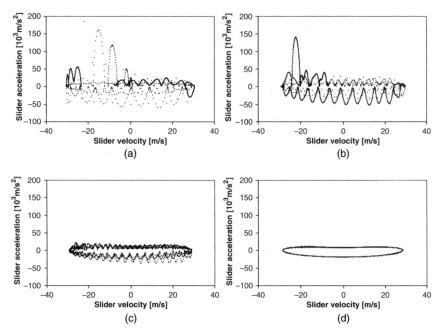

Fig. 4.18 Poincaré maps for different radial clearance sizes: (**a**) $c = 0.5$ mm; (**b**) $c = 0.2$ mm; (**c**) $c = 0.1$ mm; (**d**) $c = 0.01$ mm

response meaning that the journal and the bearing experiment a smaller number of impacts. Hence the clearance joint behavior tends to be periodic instead of nonlinear or chaotic. This conclusion is easily confirmed by the journal center trajectories and the respective Poincaré maps displayed in Fig. 4.17d and 4.18d, respectively. Figure 4.17a and b clearly shows a nonperiodic motion between the journal and bearing since it does not repeat from cycle to cycle. This is confirmed by the respective Poincaré map displayed in Fig. 4.18a and b, where the chaotic behavior can be clearly observed. This chaotic response suggests that impacts followed by some rebounds take place. Figure 4.18c shows a quasi-periodic motion, because the orbits fill up the Poincaré maps in a fully predictable manner, thus there is no sensitive dependence on the initial conditions. It is clear that, when the clearance is reduced, the dynamic response tends to be periodic or regular, which indicates that the journal follows the bearing wall. It is noteworthy that, for an ideal revolute joint, the Poincaré map is almost the same as the map shown in Fig. 4.18d, which is expected in so far as all the bodies in the system exhibit a periodic motion.

In a way similar to the clearance size study for the crank moments, the journal center trajectories and Poincaré maps are used to quantify the behavior of the slider–crank mechanism when friction is taken into account. In the models used, the contact between the journal and the bearing is modeled by the Lankarani and Nikravesh force model, given by (3.9), together with the modified Coulomb friction law, given by (3.16). The radial clearance size is equal to 0.5 mm and four different values

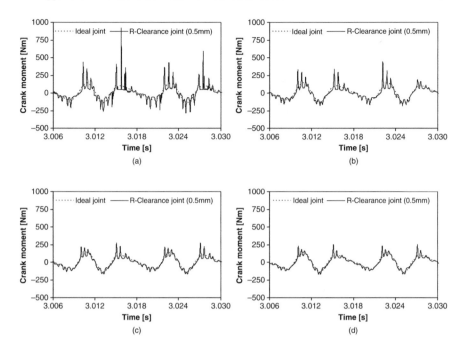

Fig. 4.19 Crank moment for different values of friction coefficient: (**a**) $c_f = 0.01$; (**b**) $c_f = 0.03$; (**c**) $c_f = 0.05$; (**d**) $c_f = 0.1$

for friction coefficient are used, namely, 0.01, 0.03, 0.05 and 0.1. Figure 4.19 depicts the crank moments for different friction coefficient values. The journal center trajectories and the corresponding Poincaré maps are shown in Figs. 4.20 and 4.21.

In general, the effect of the friction is to reduce the peaks of the force values due to the impact between the journal and the bearing. This can be observed in the crank moment plots, displayed in Fig. 4.19a–d. Figure 4.20a–d shows that the path of the journal center is characterized by a continuous contact, i.e., the journal follows the bearing wall all the time when the friction coefficient is increased. By observing Figures 4.19–4.21 it is clear that, when the friction coefficient increases, the dynamic response of the system tends to be periodic and closer to that of the system with ideal joints. For a low value of the friction coefficient, the system response is chaotic since the Poincaré map has the trajectories spread, as shown in Fig. 4.21a.

In short, multibody mechanical systems with clearance joints are well known as nonlinear dynamic systems that, under certain conditions, exhibit a chaotic response. However, from the results presented here, it is found that the dynamics of the revolute clearance joint in multibody mechanical systems is sensitive not only to the clearance size but also to the friction coefficient. With a small change in one of these parameters the response of the system can shift from chaotic to periodic and vice versa.

In what follows, several models demonstrate how different contact force models may influence the behavior of the slider–crank mechanism and what their

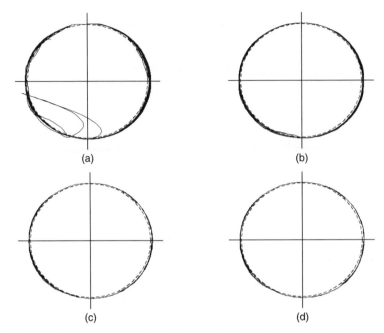

Fig. 4.20 Journal center trajectory for different values of friction coefficient: (**a**) $c_f = 0.01$; (**b**) $c_f = 0.03$; (**c**) $c_f = 0.05$; (**d**) $c_f = 0.1$

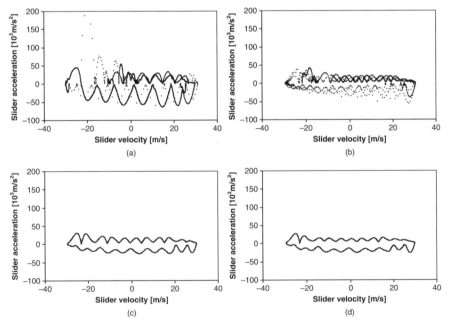

Fig. 4.21 Poincaré maps for different values of friction coefficient: (**a**) $c_f = 0.01$; (**b**) $c_f = 0.03$; (**c**) $c_f = 0.05$; (**d**) $c_f = 0.1$

consequences are in terms of the contact force and crank reaction moment. The contact force models for both spherical and cylindrical contact surfaces, presented in Chap. 3, are used here. The impact is treated as being frictionless and the radial clearance size is set to be 0.5 mm. The contact force for two full crank rotations is used to illustrate the behavior of the slider–crank mechanism, when different contact force models are applied. The contact forces corresponding to each model are pictured in Fig. 4.22. In addition to the contact force, the driving crank moment necessary to maintain a constant crank angular velocity for the different contact models and the journal center trajectories are presented in Figs. 4.23 and 4.24, respectively.

From Fig. 4.22, it is clear that for all pure elastic contact models the level of contact force is higher when compared to the continuous contact force model by

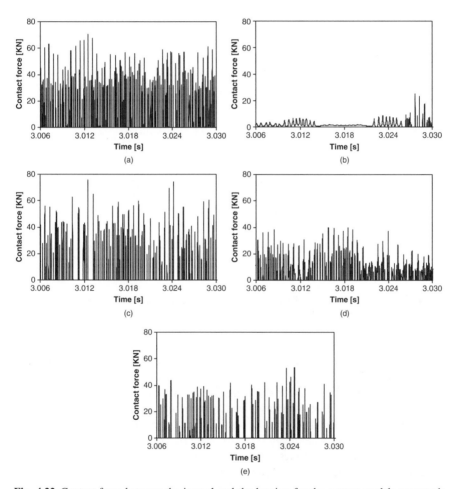

Fig. 4.22 Contact force between the journal and the bearing for the contact models presented: (**a**) Hertz contact law; (**b**) Lankarani and Nikravesh model; (**c**) Dubowsky and Freudenstein model; (**d**) Goldsmith model; (**e**) ESDU 78035 model

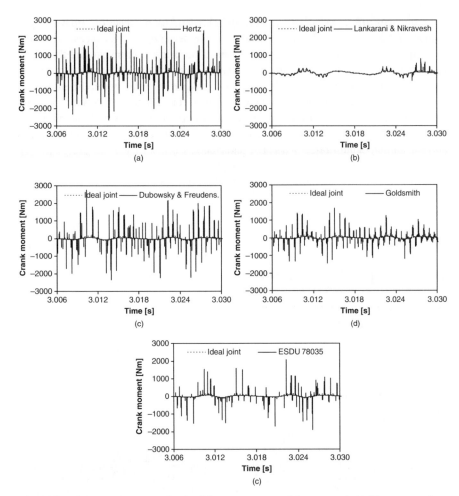

Fig. 4.23 Driving crank moment for the different contact models presented: (**a**) Hertz contact law; (**b**) Lankarani and Nikravesh model; (**c**) Dubowsky and Freudenstein model; (**d**) Goldsmith model; (**e**) ESDU 78035 model

Lankarani and Nikravesh (1990) with a restitution coefficient $c_e = 0.9$. A similar conclusion can be drawn from Fig. 4.23a–e, since the impact forces are propagated through the rigid bodies of the slider–crank mechanism. By inspecting Fig. 4.24a–e it is observed that the contact models, which do not include energy dissipation, have short periods of contact between the journal and the bearing and long free flight periods. The contact model given by (3.9), which accounts for energy dissipation, presents long periods of contact between the journal and the bearing.

In Fig. 4.24a–e, the journal trajectories are presented by continuous lines that connect points. A point is plotted for each integration time step and represented by a marker, the relative penetration depth being visible by points outside the clearance circle. The point density is very high when the journal contacts the bearing wall,

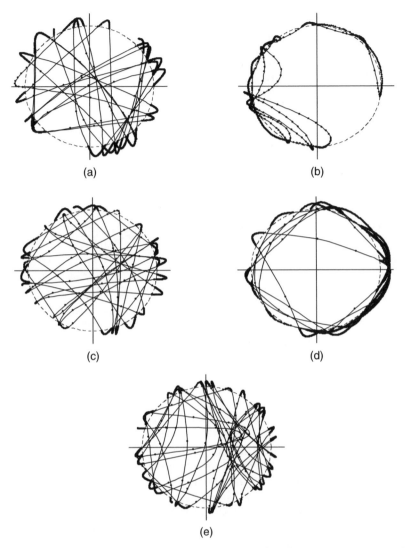

Fig. 4.24 Journal center trajectory for the different contact models: (**a**) Hertz contact law; (**b**) Lankarani and Nikravesh model; (**c**) Dubowsky and Freudenstein model; (**d**) Goldsmith model; (**e**) ESDU 78035 model

which means that the step size of the integration algorithm is small. When the journal is in free flight motion, the time step is automatically increased by the integration algorithm and, consequently, the points plotted in Fig. 4.24a–e are further apart. This shows the importance of using a varying time-step integration scheme for the dynamic analysis of systems that involve contact and impact.

Figure 4.25a–d depicts the influence of the coefficient of restitution on the crank moment when the continuous contact force model proposed by Lankarani and

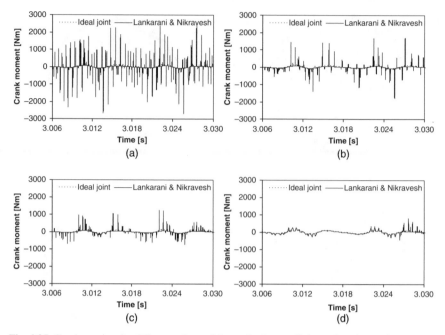

Fig. 4.25 Crank moment for different values of the restitution coefficient when the continuous contact force model proposed by Lankarani and Nikravesh (1990) is used: (**a**) $c_e = 1.00$; (**b**) $c_e = 0.99$; (**c**) $c_e = 0.95$; (**d**) $c_e = 0.90$

Nikravesh (1990), given by (3.9), is used. This model shows a direct relationship between the contact force and coefficient of restitution that is selected to the process of energy dissipation during the impact. In the case studies presented, the radial clearance size is equal to 0.5 mm and four different coefficients of restitution are selected, namely, 1.00, 0.99, 0.95 and 0.90. In Fig. 4.25a–e, it is observed that when the coefficient of restitution decreases, the peaks of the crank moment values are reduced, indicating a higher energy dissipation during the contact.

Figure 4.26a and b shows the time variation of the contact force and the relation between the force and penetration depth for different values of the coefficient of restitution, for the first impact only. Observing Fig. 4.26b, it is important to highlight how the hysteresis loop increases when the coefficient of restitution decreases. When the restitution coefficient is the unit, which corresponds to the pure Hertz contact force law, there is no energy dissipation in the contact process. This result is evident in the force-penetration depth relation of Fig. 4.26b, which does not present an hysteresis loop.

Some important conclusions can be drawn from the study presented here. Among the spherical shaped contact areas, the linear Kelvin–Voigt contact model does not represent the overall nonlinear nature of impact. The Hertz relation, besides its nonlinearity, does not account for the energy dissipation during the impact process. Therefore the Hertz relation along with the modification to represent the energy

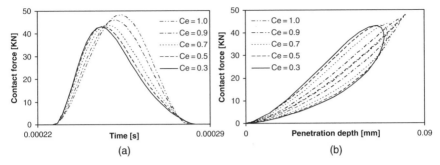

Fig. 4.26 Influence of the coefficient of restitution for the Lankarani and Nikravesh continuous contact force model: (**a**) contact force versus time; (**b**) force–penetration depth relation

dissipation in the form of internal damping is advantageous for modeling contact forces in a multibody system. The cylindrical models are expressed by nonlinear and implicit functions of penetration depth and they require a numerical iterative procedure to solve them, if no approximation is to be used.

The different continuous contact force models which use an elastic contact theory lead to comparable results in terms of contact forces, crank moments and journal trajectories. However, when dissipation is allowed to take place, the peaks of the crank moments that are required to drive the mechanism with a constant angular velocity are much lower than those observed for the elastic models. This observation is consistent with the comparisons of the flight trajectories observed for the different models. It was observed that the energy dissipation of the continuous contact model, proposed by Lankarani and Nikravesh (1990), results in long periods of time when the journal seats in the bearing, thus predicting a much smoother dynamic response of the system. Furthermore because the restitution coefficient plays a role in the control of the energy dissipation this model can represent a much broader number of contact conditions.

4.5 Application 2: Slider–Crank with Translational Clearance Joint

In order to examine the consequences of the formulation developed for the translational clearance joint, the planar slider–crank mechanism, discussed in the previous section, is considered again as a numerical example here. All joints are ideal except for the translation joint that has a clearance, as shown in Fig. 4.27. The translational clearance joint is composed of a guide and a slider. This joint has a finite clearance, which is constant along the length of the slider.

It is assumed that the crank is driven at a constant angular velocity equal to 5000 rpm, maintained by varying the input torque. Initially the slider is at the same distance from the upper and lower guide surfaces and the initial velocities and positions

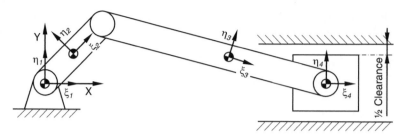

Fig. 4.27 Slider–crank mechanism with a translational clearance joint

are those used in Sect. 4.4. Table 4.3 shows the parameters used in the simulation of this demonstrative example.

The dynamic performance of the slider–crank mechanism study case is demonstrated through the time plots of the slider velocity and acceleration and the moment that acts on the crank, which are presented in Figure 4.28. Additionally the slider trajectories inside the guide are represented in Fig. 4.28 in a nondimensional plot. Results for two full crank rotations are given in Fig. 4.28a–d. The impact between the slider and guide surfaces is assumed to be frictionless, the Lankarani and Nikravesh contact force model expressed by (3.9) being used to compute the normal forces. For convenience, a small radius of curvature at each slider corner is considered, in order to calculate the equivalent generalized stiffness given by (3.5).

The slider velocities and acceleration, presented in Fig. 4.28a and b, clearly show the influence of the clearance in the kinematics of the translation joint. The slider velocity diagram is smooth and close to the ideal joint simulation. The smooth changes in the velocity also indicate that the slider and guide surfaces are in permanent contact for long periods. Some sudden changes in the velocity are due to the impacts between the slider and guide surfaces. These impacts are visible in the acceleration diagram by high values in the form of peaks in the response. Since the bodies of the slider–crank mechanism are rigid, the impact forces are propagated from the slider to the crank, leading to visible high peaks in the crank moment diagram shown in Fig. 4.28c.

The dimensionless slider trajectories are shown in Fig. 4.28d. There, the different types of motion between the slider and guide observed are associated with the different guide–slider configurations, that is, no contact, contact-impact followed by rebound and permanent contact. The dimensionless X-slider motion varies from

Table 4.3 Numerical parameters used in the dynamic simulation of the slider–crank mechanism with a translational clearance joint

Clearance size	0.5 mm	Young's modulus	207 GPa
Slider length	50.0 mm	Poisson's ratio	0.3
Slider width	50.0 mm	Baumgarte - α	5
Slider thickness	50.0 mm	Baumgarte - β	5
Corner curvature radius	1.0 mm	Integration step	10^{-5} s
Restitution coefficient	0.9	Integration tolerance	10^{-6} s

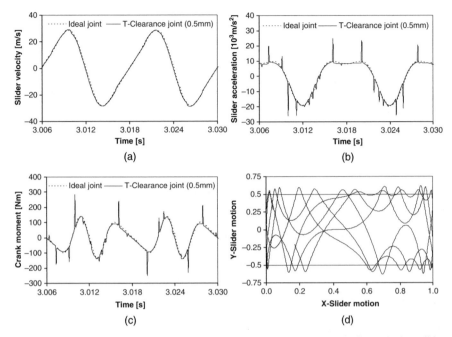

Fig. 4.28 (**a**) Slider velocity; (**b**) slider acceleration; (**c**) crank moment; (**d**) dimensionless slider trajectories inside the guide

0 to 1, which corresponds to the low and top dead ends, respectively. The dimensionless Y-slider motion higher than 0.5 corresponds to the case in which the slider and the upper guide surface are in contact, whereas the dimensionless Y-slider motion lower than −0.5 corresponds to the case in which the contact takes place between the slider and the lower guide surface. The horizontal lines in the slider path diagrams represent the geometric limits for contact situations between the slider and guides surfaces.

In order to understand the influence of the clearance size in the dynamic behavior of the slider–crank mechanism, the driving crank moment is plotted in Fig. 4.29a–d for simulations where the clearance in the translational clearance joint varies from 0.01 to 0.5 mm. In addition to the crank moment, the respective slider trajectories and the Poincaré maps are presented in Figs. 4.30 and 4.31. Again the slider acceleration and slider velocity are chosen to build the Poincaré maps.

From Fig. 4.29 it is evident that when the clearance size is small the crank moment peaks are lower and the dynamic response tends to be closer to the ideal translation joint case. This suggests that the periods of permanent contact between the slider and guide surfaces are longer and, hence, the slider and guide experiment fewer impacts. This observation can be confirmed by the slider trajectories and Poincaré maps provided in Figs. 4.30 and 4.31, respectively. When the clearance size is reduced, the system response changes from chaotic, as displayed in

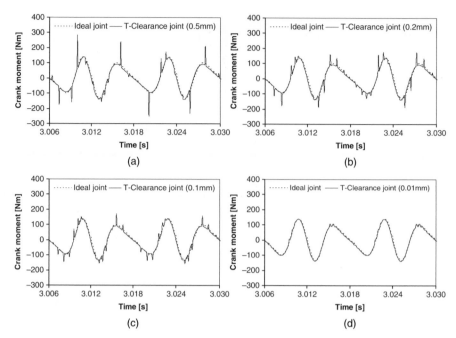

Fig. 4.29 Driving crank moment for different clearance sizes in the translational clearance joint:
(**a**) $c = 0.5$ mm; (**b**) $c = 0.2$ mm; (**c**) $c = 0.1$ mm; (**d**) $c = 0.01$ mm

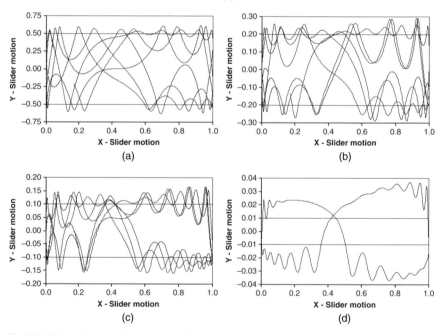

Fig. 4.30 Dimensionless slider path for different clearance sizes in the translational clearance joint:
(**a**) $c = 0.5$ mm; (**b**) $c = 0.2$ mm; (**c**) $c = 0.1$ mm; (**d**) $c = 0.01$ mm

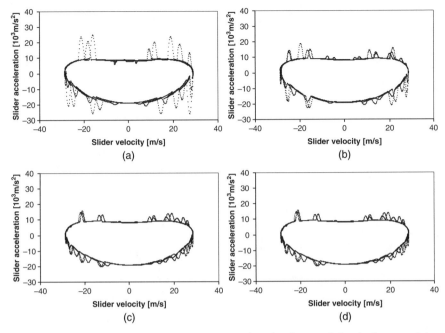

Fig. 4.31 Poincaré maps for different clearance sizes in the translational clearance joint: (a) $c = 0.5$ mm; (b) $c = 0.2$ mm; (c) $c = 0.1$ mm; (d) $c = 0.01$ mm

Fig. 4.30a, to periodic or regular, as observed in Fig. 4.31d. This feature can be useful in the evaluation of the acceptable range for clearance, in any type of construction where this type of joints is applied.

The effect of the friction phenomenon on the dynamic performance of the translational clearance joint is also studied. Again the driving crank moment, the slider path and Poincaré maps are used to quantify the dynamic response of the slider–crank mechanism and represented in Figs. 4.32–4.34. The value for the clearance size is equal to 0.5 mm in all simulations, and four different values for the friction coefficient are used, namely, 0.01, 0.03, 0.05 and 0.1.

In Fig. 4.32, it is observed that the reaction, or driving crank moment necessary to maintain constant crank angular velocity, does not relate directly to the friction coefficient value, that is, when the friction coefficient increases the peak values of the crank moment do not show tendency to increase or decrease. Analyzing the slider trajectories, plotted in Figs. 4.33a–d, and the corresponding Poincaré maps, shown in Fig. 4.34a–d, it is observed that the influence of the friction coefficient in global system response is not significant, conversely to what happens in the case of the revolute clearance joint, shown in Figs. 4.19–4.21.

The influence of employing different contact force models on the global slider–crank behavior is also analyzed in this work. Figures 4.35 and 4.36 show the contact forces and driving crank moments for the contact force models given by (3.9) and (4.26), respectively, that is, the nonlinear force model proposed by Lankarani and

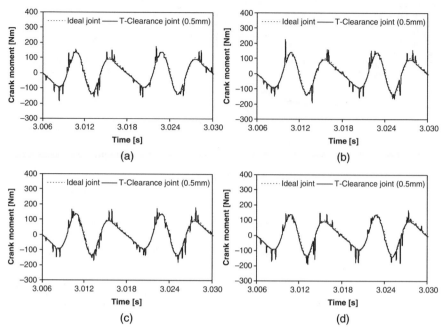

Fig. 4.32 Driving crank moment for different values of the friction coefficient in the translational clearance joint: (**a**) $c_f = 0.01$; (**b**) $c_f = 0.03$; (**c**) $c_f = 0.05$; (**d**) $c_f = 0.1$

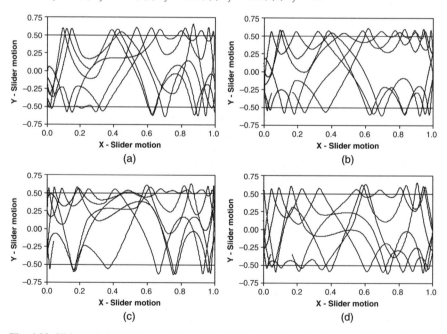

Fig. 4.33 Slider path for different values of the friction coefficient in the translational clearance joint: (**a**) $c_f = 0.01$; (**b**) $c_f = 0.03$; (**c**) $c_f = 0.05$; (**d**) $c_f = 0.1$

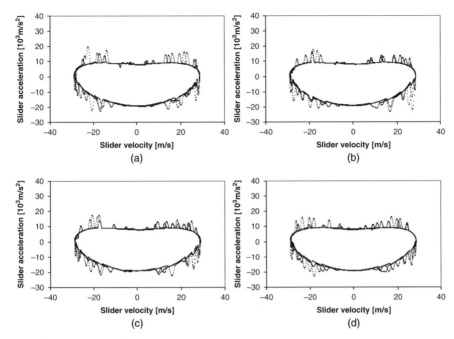

Fig. 4.34 Poincaré maps for different friction coefficients in the translational clearance joint: (a) $c_f = 0.01$; (b) $c_f = 0.03$; (c) $c_f = 0.05$; (d) $c_f = 0.1$

Nikravesh (1990), using a restitution coefficient of $c_e = 0.9$, and the linear force model for two plane surfaces presented by Lankarani (1988). In the case of the linear contact model for two plane surfaces, the average penetration is used to evaluate the magnitude of the contact force. This force is then applied at the geometric center of the penetration area, as schematically shown in Fig. 4.11a.

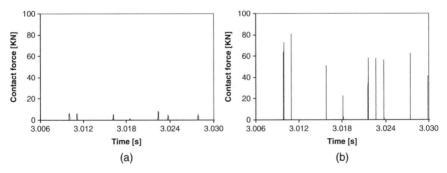

Fig. 4.35 Contact force between the slider and guide surface: (a) Lankarani and Nikravesh model; (b) linear contact model for two plane surfaces

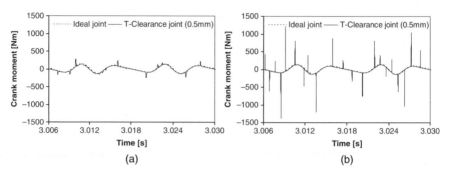

Fig. 4.36 Driving crank moment: (**a**) Lankarani and Nikravesh model; (**b**) linear contact model for two plane surfaces

4.6 Summary

A comprehensive approach to the modeling of unlubricated clearance joints in planar multibody systems has been presented in this chapter, with specialization for the revolute and translational joints. In the process, different contact models have been revised in face of their suitability to represent the impact between the bodies joined by these joints. The methodologies proposed have been exemplified, through the application to the dynamic study of a slider–crank mechanism with revolute and translational clearance joints.

A critical aspect, in the precise prediction of the peak forces, is the proper selection of an appropriate contact model. From the comparison between the cylindrical contact force models and spherical contact force models, one can conclude that the spherical and cylindrical force–displacement relations are reasonably close. Furthermore the straightforward force–penetration relation proposed by Lankarani and Nikravesh (1990) is largely used for mechanical contacts owing to its simplicity and easiness of implementation in a computational program and also because this is the only model that accounts for energy dissipation during the impact process. A modified Coulomb's friction law was used to model the friction phenomenon; one merit of this modified model is that it improves the numerical stability of the integration algorithm.

How the solution strategy of the contact problem associated with the modeling of joints with clearances is sensitive to the procedure used to detect contact was discussed. In the sequel of the techniques proposed, a numerical strategy that takes advantage of the use of a variable time-step integration algorithm has been proposed to handle the identification of the start of contact and to proceed afterwards.

The dynamic response of the slider–crank mechanism with clearance joints, in terms of Poincaré maps, shows that periodic and nonperiodic responses can occur. Poincaré maps play a key role in representing the global behavior of dynamical systems. For the radial clearance and friction coefficient values used in this work the system response exhibits both periodic and chaotic responses. When the clearance is reduced, the dynamic response changes from chaotic to periodic or regular behavior.

In fact, most of the mechanical systems present some inherent nonlinearity even when modeled with ideal joints. The chaotic behavior of the mechanical system may be eliminated with suitable design and/or parameter changes of mechanical system.

For the case of revolute clearance joint analysis, the overall results are corroborated by published works on this field, for cases that include dry contact force models (Ravn 1998, Schwab 2002). The results for translational clearance joints in mechanical systems seem quite acceptable, but they cannot be compared with other results since this issue has not been addressed before in the literature.

The planar slider–crank mechanism with a translational clearance joint was used as a numerical example to illustrate the methodology proposed. In general, the dynamic response of the slider–crank mechanism presents some peaks, due to the impact between the slider and the guide, namely in what concerns the accelerations and reaction moments. It was observed that all curves for the kinematic and kinetic variables are similar to those obtained with ideal joints, with the exception of the peaks especially visible for the forces and accelerations. These peaks have been clearly associated to the existence of the clearances and to their magnitude. The relative motion between the guide and slider showed a very high nonlinearity, or even chaotic behavior, when a translational clearance is included. When the clearance size is reduced, the system's response becomes closer to the case for ideal joints. Furthermore the dynamic behavior of the slider–crank model tends to be periodic or regular. This feature can be useful in calculations of an acceptable range for the clearance, during the design process.

The conclusions drawn from the results presented in this chapter must be considered in the light of the assumptions identified in the formulation of the motion's equation, namely the assumption of rigidity for the bodies, the lack of joints' flexibility and the nonexistence of lubrication effects. In this chapter, contact between the surfaces that constitute the clearance joints was assumed to be dry, i.e., without any interposition fluid layer. Consequently contact-impact forces and friction forces were the only loads on the contacting bodies when the physical contact was detected between the surfaces. In the engineering design of machines and mechanisms, journal–bearings are usually designed to operate with some lubricant. Lubricated journal–bearings are designed so that when the maximum load is applied, the journal and bearing do not come into contact. The main reason for designing journal–bearings in this way is to reduce the friction and extend the lifetime of mechanical systems. The issue of lubricated journal–bearings in mechanical systems is presented and discussed in the next chapter.

References

Ambrósio JAC (2002) Impact of rigid and flexible multibody systems: deformation description and contact models. Virtual nonlinear multibody systems, NATO Advanced Study Institute, Prague, Czech Republic, June 23–July 3, edited by W Schiehlen and M Valásek, Vol. II, pp. 15–33.
Baker GJ, Gollub JP (1990) Chaotic dynamics—an introduction. Cambridge University Press, Cambridge, United Kingdom.

Bauchau OA, Rodriguez J (2002) Modelling of joints with clearance in flexible multibody systems. International Journal of Solids and Structures 39:41–63.

Bengisu MT, Hidayetoglu T, Akay A (1986) A theoretical and experimental investigation of contact loss in the clearances of a four-bar mechanism. Journal of Mechanisms, Transmissions, and Automation in Design 108:237–244.

Dubowsky S (1974) On predicting the dynamic effects of clearances in planar mechanisms. Journal of Engineering for Industry, Series B 96(1):317–323.

Dubowsky S, Freudenstein F (1971a) Dynamic analysis of mechanical systems with clearances, part 1: formulation of dynamic model. Journal of Engineering for Industry, Series B 93(1):305–309.

Dubowsky S, Freudenstein F (1971b) Dynamic analysis of mechanical systems with clearances, part 2: dynamic response. Journal of Engineering for Industry, Series B 93(1):310–316.

Earles SWE, Wu CLS (1975) Predicting the occurrence of contact loss and impact at the bearing from a zero-clearance analysis. Proceedings of IFToMM fourth world congress on the theory of machines and mechanisms, Newcastle Upon Tyne, England, pp. 1013–1018.

Farahanchi F, Shaw, SW (1994) Chaotic and periodic dynamics of a slider crank mechanism with slider clearance. Journal of Sound and Vibration 177(3):307–324.

Flores P, Ambrósio J (2004) Revolute joints with clearance in multibody systems. Computers and Structures, Special Issue: Computational Mechanics in Portugal 82(17–18):1359–1369.

Flores P, Ambrósio J, Claro JCP, Lankarani HM, Koshy CS (2006) A study on dynamics of mechanical systems including joints with clearance and lubrication. Mechanism and Machine Theory 41(3):247–261.

Hertz H (1896) On the contact of solids—on the contact of rigid elastic solids and on hardness. Miscellaneous papers (Translated by DE Jones and GA Schott), pp. 146–183. Macmillan, London, England.

Lankarani HM (1988) Canonical equations of motion and estimation of parameters in the analysis of impact problems. Ph.D. Dissertation, University of Arizona, Tucson, AZ.

Lankarani HM, Nikravesh PE (1990) A contact force model with hysteresis damping for impact analysis of multibody systems. Journal of Mechanical Design 112:369–376.

Mansour WM, Townsend MA (1975) Impact spectra and intensities for high-speed mechanisms. Journal of Engineering for Industry, Series B 97(2):347–353.

Miedema B, Mansour WM (1976) Mechanical joints with clearance: a three mode model. Journal of Engineering for Industry 98(4):1319–1323.

Nikravesh PE (1988) Computer-aided analysis of mechanical systems. Prentice Hall, Englewood Cliffs, NJ.

Ravn P (1998) A continuous analysis method for planar multibody systems with joint clearance. Multibody System Dynamics 2:1–24.

Schwab AL (2002) Dynamics of flexible multibody systems, small vibrations superimposed on a general rigid body motion. Ph.D. Dissertation, Delft University of Technology, Netherlands.

Shabana AA (1989) Dynamics of multibody systems. Wiley, New York.

Thümmel T, Funk K (1999) Multibody modelling of linkage mechanisms including friction, clearance and impact. Proceedings of tenth world congress on the theory of machine and mechanisms, Oulu University, Finland, Vol. 4, pp. 1375–1386.

Timoshenko SP, Goodier JN (1970) Theory of elasticity. McGraw-Hill, New York.

Tomsen JJ (1997) Vibrations and stability—order and chaos, McGraw-Hill, New York.

Townsend MA, Mansour WM (1975) A pendulating model for mechanisms with clearances in the revolutes. Journal of Engineering for Industry, Series B 97(2):354–358.

Wiggins S (1990) Introduction to applied nonlinear dynamical systems and chaos. Springer Berlin Heidelberg New York.

Wilson R, Fawcett JN (1974) Dynamics of slider–crank mechanism with clearance in the sliding bearing. Mechanism and Machine Theory 9:61–80.

Wu CLS, Earles SWE (1977) A determination of contact-loss at a bearing of a linkage mechanism. Journal of Engineering for Industry, Series B 99(2):375–380.

Zukas JA, Nicholas T, Greszczuk LB, Curran DR (1982) Impact dynamics. Wiley, New York.

Chapter 5
Lubricated Joints for Mechanical Systems

In most machines and mechanisms, the joints are designed to operate with some lubricant fluid. The high pressures generated in the lubricant fluid act to keep the journal and the bearing apart. Moreover the thin film formed by lubricant reduces friction and wear, provides load-carrying capacity and adds damping to dissipate undesirable mechanical vibrations (Hamrock 1994, Frêne et al. 1997). Therefore the proper description of lubricated revolute joints, the so-called journal–bearings, in multibody systems is required to achieve better models and hence an improved understanding of the dynamic performance of machines. This aspect gains paramount importance due to the demand for the proper design of journal–bearings in many industrial applications (Ravn 1998, Bauchau and Rodriguez 2002). In the dynamic analysis of journal–bearings, the hydrodynamic forces, which include both squeeze and wedge effects, generated by the lubricant fluid, oppose the journal motion. It should be mentioned that the methodology presented here uses the superposition principle for load capacity due to the wedge effect entrainment and squeeze-film effect separately (Hamrock 1994). The hydrodynamic forces are obtained by integrating the pressure distribution evaluated with the aid of Reynolds' equation written for the dynamic regime (Pinkus and Sternlicht 1961). The hydrodynamic forces are nonlinear functions of the journal center position and of its velocity with reference to the bearing center. In the dynamic regime of a journal–bearing, the journal center has an orbit situated within a circle radius which is equal to the radial clearance. Thus a lubricated revolute joint does not impose kinematic constraints like an ideal revolute joint but instead it deals with force constraints as for the dry clearance joints. In a simple way, the hydrodynamic forces built up by the lubricant fluid are evaluated from the state of variable of the system and included into the equations of motion of the mechanical system (Nikravesh 1988). For dynamically loaded journal–bearings the classic analysis problem consists in predicting the motion of the journal center under arbitrary and known loading (Boker 1965, Goenka 1984). However, in the present work the time variable parameters are known from the dynamic system's configuration and the instantaneous force on the journal–bearing are evaluated afterwards.

In this chapter, a general methodology for modeling lubricated revolute joints in multibody mechanical systems is presented and discussed. A simple journal–bearing

P. Flores et al., *Kinematics and Dynamics of Multibody Systems with Imperfect Joints.*   101
© Springer 2008

subjected to a constant and unidirectional external load and a simple planar slider–crank mechanism, in which a lubricated revolute joint in the gudgeon-pin exists, are used as numerical applications in order to demonstrate the assumptions and procedures adopted.

5.1 General Issues in Tribology

In physical joints, clearance, friction and impact are always present, mainly when there is no fluid lubricant. These phenomena can significantly change the dynamic response of the mechanical systems in so far as the impact causes noise, increases the level of vibrations, reduces the fatigue life of the components and results in loss of precision. When the clearance joints are dry, i.e., without lubricant, contact and friction forces are the only effects present, representing the physical contact detected between the surfaces (Flores and Ambrósio 2004). However, in most the mechanical systems, the joints are designed to operate with some lubricant fluid. It is known that the use of lubricant in revolute joints is an effective way of ensuring better performance of the mechanical systems. In fact, lubricants are widely used in thin fluid film journal–bearing elements to reduce friction and wear, provide load capacity and add damping to dissipate undesirable mechanical vibrations.

Journal–bearing is a circular shaft, designated as journal, rotating in a circular bush, designated as bearing. The space between the two elements is filled with a lubricant. Under an applied load, the journal center is displaced from the bearing center and the lubricant is forced into the convergence clearance space causing a build-up of pressure. The high pressures generated in the lubricant film act to keep the journal and the bearing surfaces apart. It is well known from fluid mechanics that a necessary condition to develop pressure in a thin film is that the gradient and the slope of the velocity profile vary across the thickness of the film (Frêne et al. 1997).

Shown in Fig. 5.1 is the basic journal–bearing geometry. The length of the bearing is L and it has a diameter D. The difference between the radii of the bearing and the journal is the radial clearance c ($c = R_B - R_J$). In general, both the journal and the bearing may rotate nonuniformly, and the applied load may vary in both magnitude and direction.

In the region of the local converging film thickness, the hydrodynamic pressure rises to a maximum value and then decreases to ambient values at the side and trailing edges of the thin film. In zones where the film thickness locally increases, the fluid pressure may drop to ambient or below its vapor pressure leading to the release of dissolved gases within the lubricant vaporization causing film rupture. The phenomenon of film rupture is known as lubricant cavitation, and the effects on the performance and stability of journal–bearings are reasonably well understood and documented in the literature (Dowson and Taylor 1979, Elrod 1981, Woods and Brewe 1989, Mistry et al. 1997). The performance of journal–bearings considering lubricant supply conditions has been studied theoretically and experimentally by Miranda (1983), Claro (1994) and Costa (2000).

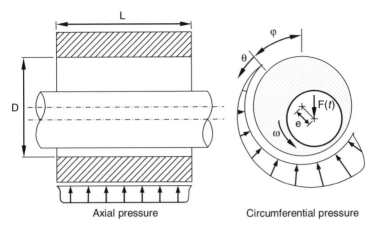

Axial pressure   Circumferential pressure

Fig. 5.1 Basic journal–bearing geometry

Journal–bearings are used in many important operating situations, in which the loads vary in both magnitude and direction, often cyclically. Examples include reciprocating machinery such as internal combustion engines, compressors, out of balance rotating machinery such as turbine rotors and other industrial processing machinery. The hydrodynamic fluid film developed in the journal–bearings exhibits damping which plays a very important role in the stability of the mechanical elements. In order to study the performance of such journal–bearings, it is necessary to determine the loads and their variation with time. In dynamically loaded journal–bearings, the eccentricity and the attitude angle vary through the loading cycle, and special care must be taken to ensure that the combination of load and speed rotation does not lead to a dangerous small minimum film thickness.

Lubricated joints are designed so that even when the maximum load is applied, the journal and bearing do not come in contact. One of the main reasons for designing journal–bearings in this way is to reduce friction and extend their lifetime. Consequently proper modeling of lubricated revolute joints in multibody mechanical systems is required to achieve a better understanding of the dynamic performance of the machines. This aspect plays a crucial role owing to the demand for proper design of the journal–bearings in many industrial applications.

5.2 Dynamic Characteristics of Journal–Bearings

If the journal and bearing have relative angular velocities with respect to each other, the amount of eccentricity adjusts itself until the pressure generated in the converging lubricating film balances the external loads. The pressure generated, and hence the load capacity of the journal–bearing, depends on the journal eccentricity, the relative angular velocity, the effective viscosity of the fluid lubricant and the journal–bearing geometry and clearance. There are two different actions of pressure

Fig. 5.2 (a) Squeeze-film
action; (b) wedge-film action

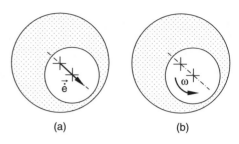

(a)   (b)

generation in journal–bearings, namely wedge and squeeze actions, as shown in
Fig. 5.2. The squeeze action relates the radial journal motion to the generation of
load capacity pressure in the lubricant film, whilst the wedge action deals with the
relation between relative rotational velocity of the journal and bearing ability to
produce such pressure.

When only the squeeze action of the lubricant is considered, assuming a null or
low relative rotational velocity and, hence, absence of relative tangential velocity,
the journal load and the fluid reaction force are considered to have the same line of
action, which is collinear with the center lines. However, in the more general case,
in the presence of high angular velocities, they do not have the same line of action
because of the wedge effect. When relative angular velocities are large, the simple
squeeze approach is not valid and the general Reynolds' equation has to be used
(Hamrock 1994, Frêne et al. 1997).

In general, mechanical systems demand journal–bearings in which the load varies
in both magnitude and direction, which results in dynamically loaded journal–
bearings. Figure 5.3 shows the cross-section of a smooth dynamically loaded
journal–bearing. When the load acting on the journal–bearing is not constant in
direction and/or module, the journal center describes an orbit within the bearing boundaries. Typical examples of dynamically loaded journal–bearings include

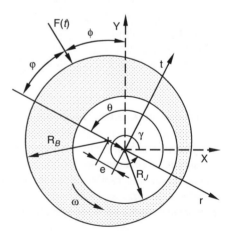

Fig. 5.3 Cross-section of a
smooth dynamically loaded
journal–bearing

the crankshaft bearings in combustion engines and high-speed turbine bearings supporting dynamic loads caused by unbalanced rotors.

In mechanical systems, a lubricated revolute joint does not produce any kinematic constraint. Instead it acts in a similar way to a force element producing time-dependent forces. Thus it deals with the so-called force constraints. For dynamically loaded journal–bearings the classic tribology analysis problem consists in predicting the motion of the journal center under arbitrary and known loads, using, for instance, the mobility method (Boker 1965, Goenka 1984). Conversely, in the work now presented, the time-variable parameters are known from the dynamic analysis and the instantaneous forces on the journal–bearing are calculated. In a simple way, the forces built up by the lubricant fluid are evaluated from the state of variables of the system and included into the equations of motion of the mechanical system as external generalized forces.

5.3 Hydrodynamic Forces in Dynamic Journal–Bearings

Theory of lubrication for dynamically loaded journal–bearings is mathematically complex, and the solution of the governing differential equations is based on many simplifying premises. The main basic principles, terminology and theoretical background are well discussed in the thematic literature, such as Hamrock (1994), Frêne et al. (1997), amongst others. Pinkus and Sternlicht (1961) present a detailed derivation of the Reynolds' equation, in which the forces developed by the fluid film pressure field are evaluated. The Reynolds' equation contains viscosity, density and film thickness as parameters. The full general form of the isothermal Reynolds' equation for a dynamically loaded journal–bearing is written as

$$\frac{\partial}{\partial X}\left(\frac{h^3}{\mu}\frac{\partial p}{\partial X}\right) + \frac{\partial}{\partial Z}\left(\frac{h^3}{\mu}\frac{\partial p}{\partial Z}\right) = 6U\frac{\partial h}{\partial X} + 12\frac{dh}{dt} \tag{5.1}$$

where X is the radial direction, Z is the axial direction, μ is the dynamic fluid viscosity, h denotes the film thickness and p is the pressure. The two terms on the right-hand side of (5.1) represent the two different effects of pressure generation on the lubricant film, i.e., wedge and squeeze actions, respectively.

It is known that (5.1) is a nonhomogeneous partial differential of the elliptical type. The exact solution of the Reynolds' equation is difficult to obtain and, in general, requires a considerable numerical effort. However, it is possible to solve the equation analytically by setting to zero either the first or second term on the left-hand side. These solutions correspond to those for infinitely short and infinitely long journal–bearings, respectively.

Dubois and Ocvirk (1953) consider a journal–bearing where the pressure gradient around the circumference is very small when compared with those along the length. This assumption is valid for length-to-diameter (L/D) ratios up to 0.5. Hence the Reynolds' equation for an infinitely short journal–bearing can be written as

$$\frac{\partial}{\partial Z}\left(\frac{h^3}{\mu}\frac{\partial p}{\partial Z}\right) = 6U\frac{\partial h}{\partial X} + 12\frac{dh}{dt} \tag{5.2}$$

When the relative pressure is zero at journal–bearing ends the pressure in the fluid film is given by (Frêne et al. 1997)

$$p(\theta, Z) = -\frac{3\mu}{h^3}\left(\frac{L^2}{4} - Z^2\right)\left((\omega - 2\dot{\gamma})\frac{\partial h}{\partial \theta} + 2\dot{e}\cos\theta\right) \tag{5.3}$$

where θ is the angular coordinate, Z is the axial direction, L represents the journal–bearing length, μ is the dynamic fluid viscosity, h denotes the film thickness and ω is the relative angular velocity between the journal and bearing. The dot in the above expression denotes the time derivative of the corresponding parameter.

For an infinitely long journal–bearing a constant fluid pressure and negligible leakage in the axial direction are assumed. In many cases, it is possible to treat a journal–bearing as infinitely long and consider only its middle point. This solution was first derived by Sommerfeld (1904) and is valid for length-to-diameter (L/D) ratios greater than 2. Thus the Reynolds' equation for an infinitely long journal–bearing is

$$\frac{\partial}{\partial X}\left(\frac{h^3}{\mu}\frac{\partial p}{\partial X}\right) = 6U\frac{\partial h}{\partial X} + 12\frac{dh}{dt} \tag{5.4}$$

And the pressure distribution in the fluid is given by (Frêne et al. 1997)

$$p = 6\mu\left(\frac{R_J}{c}\right)^2\left\{\frac{(\omega - 2\dot{\gamma})(2 + \varepsilon\cos\theta)\varepsilon\sin\theta}{(2 + \varepsilon^2)(1 + \varepsilon\cos\theta)^2} + \frac{\dot{\varepsilon}}{\varepsilon}\left[\frac{1}{(1 + \varepsilon\cos\theta)^2} - \frac{1}{(1 + \varepsilon)^2}\right]\right\} \tag{5.5}$$

where R_J is the journal radius, c is the radial clearance and ε is the eccentricity ratio.

It is convenient to determine the force components of the resultant pressure in directions tangential and perpendicular to the line of centers. These force components can be obtained by integrating the pressure field either in the entire domain 2π or half-domain π. In the later case, the pressure field is integrated only over the positive part by setting the pressure in the remaining portion equal to zero. These boundary conditions, associated with the pressure field, correspond to Sommerfeld's and Gümbel's boundary conditions (Fig. 5.4).

The Sommerfeld's boundary conditions, complete or full film, do not take into account the cavitation phenomenon and, consequently, contemplate the existence of negative pressures for the region $\pi < \theta < 2\pi$. This case is not realised in many applications due to the fluid incapacity to sustain significant sub-ambient pressures.

The Gümbel's conditions, which account for the rupture film, assume the existence of a zero pressure zone for the region between π and 2π. Though the Gümbel's, or half Sommerfeld's solution, results in more realistic predictions of

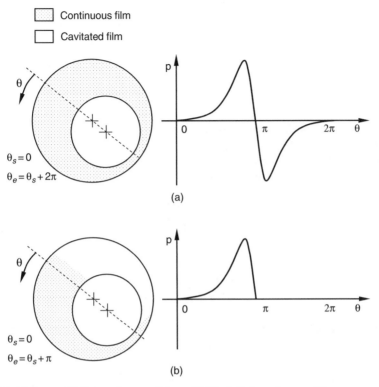

Fig. 5.4 (**a**) Sommerfeld's boundary conditions; (**b**) Gümbel's boundary conditions

the load capacity, it leads to a violation of the continuity of flow at the outlet end of the pressure curve.

For the Sommerfeld's conditions, i.e., full film, the force components of the fluid film for infinitely short journal–bearing are written as (Frêne et al. 1997)

$$F_r = -\frac{\pi\mu L^3 R_J}{c^2}\frac{\dot{\varepsilon}(1+2\varepsilon^2)}{(1-\varepsilon^2)^{5/2}} \qquad (5.6)$$

$$F_t = \frac{\pi\mu L^3 R_J}{c^2}\frac{\varepsilon(\omega-2\dot{\gamma})}{2(1-\varepsilon^2)^{3/2}} \qquad (5.7)$$

where F_r is the radial component of the force while F_t is the tangential component, as Fig. 5.3 shows.

For the Gümbel's conditions, i.e., rupture film, the force components of the fluid film for infinitely short journal–bearing can be expressed as (Frêne et al. 1997)

$$F_r = -\frac{\mu L^3 R_J}{2c^2(1-\varepsilon^2)^2}\left(\frac{\pi\dot\varepsilon(1+2\varepsilon^2)}{(1-\varepsilon^2)^{1/2}}+2\varepsilon^2(\omega-2\dot\gamma)\right) \tag{5.8}$$

$$F_t = \frac{\mu L^3 R_J \varepsilon}{2c^2(1-\varepsilon^2)^2}\left(4\dot\varepsilon+\frac{\pi}{2}(\omega-2\dot\gamma)\sqrt{1-\varepsilon^2}\right) \tag{5.9}$$

As for the case of the infinitely short journal–bearing, the complete film and the film rupture for the infinitely long journal–bearing are also distinguished. Thus, for the Sommerfeld's conditions, full film, the force components of the fluid film for infinitely long journal–bearing are written as (Frêne et al. 1997)

$$F_r = -\frac{12\pi\mu L R_J^3 \dot\varepsilon}{c^2(1-\varepsilon^2)^{3/2}} \tag{5.10}$$

$$F_t = \frac{12\pi\mu L R_J^3 \varepsilon(\omega-2\dot\gamma)}{c^2(2+\varepsilon^2)(1-\varepsilon^2)^{1/2}} \tag{5.11}$$

For the Gümbel's conditions the force components of the fluid film for infinitely long journal–bearing are written as (Frêne et al. 1997)

$$F_r = -\frac{12\mu L R_J^3}{c^2}\left(\frac{\varepsilon^2(\omega-2\dot\gamma)}{(2+\varepsilon^2)(1-\varepsilon^2)}+\frac{\dot\varepsilon}{(1-\varepsilon^2)^{3/2}}\left(\frac{\pi}{2}-\frac{8}{\pi(2+\varepsilon^2)}\right)\right) \tag{5.12}$$

$$F_t = \frac{12\mu L R_J^3}{c^2}\left(\frac{\pi\varepsilon(\omega-2\dot\gamma)}{2(2+\varepsilon^2)\sqrt{1-\varepsilon^2}}+\frac{2\varepsilon\dot\varepsilon}{(2+\varepsilon^2)(1-\varepsilon^2)}\right) \tag{5.13}$$

In fact, the main difficulty in obtaining satisfactory solutions of journal–bearing performance lies not only in solving the differential equations but also in defining adequately the boundary conditions of Reynolds' equation. In dynamically loaded journal–bearings, obtaining the force components from the integration of the Reynolds' equation only over the positive pressure regions, by setting the pressure in the remaining portions equal to zero, involves finding the zero points, i.e., the angles θ_s at which the pressure begins and θ_e when it ends. For the case of a steady-state journal–bearing, these angles are, in general, assumed to be equal to 0 and π, respectively. However, for a dynamically loaded journal–bearing these angles are time-dependent and the evaluation of the force components involves a good deal of mathematical manipulation. For details in their treatment see the work by Pinkus and Sternlicht (1961). The hydrodynamic force components, along the eccentricity direction and normal to it, are for radial velocity greater than zero, $\dot\varepsilon > 0$, given by Pinkus and Sternlicht (1961) as

$$F_r = -\frac{\mu L R_J^3}{c^2} \frac{6\dot{\varepsilon}}{(2+\varepsilon^2)(1-\varepsilon^2)^{3/2}} \left[4k\varepsilon^2 + (2+\varepsilon^2)\pi \frac{k+3}{k+3/2} \right] \qquad (5.14)$$

$$F_t = \frac{\mu L R_J^3}{c^2} \frac{6\pi\varepsilon(\omega-2\dot{\gamma})}{(2+\varepsilon^2)(1-\varepsilon^2)^{1/2}} \frac{k+3}{k+3/2} \qquad (5.15)$$

For a negative radial velocity $\dot{\varepsilon} < 0$, the force components are written as follows:

$$F_r = -\frac{\mu L R_J^3}{c^2} \frac{6\dot{\varepsilon}}{(2+\varepsilon^2)(1-\varepsilon^2)^{3/2}} \left[4k\varepsilon^2 - (2+\varepsilon^2)\pi \frac{k}{k+3/2} \right] \qquad (5.16)$$

$$F_t = \frac{\mu L R_J^3}{c^2} \frac{6\pi\varepsilon(\omega-2\dot{\gamma})}{(2+\varepsilon^2)(1-\varepsilon^2)^{1/2}} \frac{k}{k+3/2} \qquad (5.17)$$

where the parameter k is defined as

$$k^2 = (1-\varepsilon^2) \left[\left(\frac{\omega-2\dot{\gamma}}{2\dot{\varepsilon}} \right)^2 + \frac{1}{\varepsilon^2} \right] \qquad (5.18)$$

Finally the force components of the resulting pressure distribution, tangential and perpendicular to the line of centers, have to be projected onto the X and Y directions. From Fig. 5.3 it is clear that

$$F_x = F_r \cos\gamma - F_t \sin\gamma \qquad (5.19)$$

$$F_y = F_r \sin\gamma + F_t \cos\gamma \qquad (5.20)$$

Equations (5.6)–(5.18), for infinitely short and infinitely long journal–bearings, present the relation between the journal center motion and the fluid reaction force on the journal. The solution of these equations presents no problem since the journal center motion is known from the dynamic analysis of the mechanical system.

In traditional tribology analysis of journal–bearings, the external forces are known and the motion of the journal center inside the bearing boundaries is evaluated by solving the differential equations for the time-dependent variables. However, in the present work instead of knowing the applied load, the relative journal–bearing motion characteristics are known. Afterwards the fluid force, from the pressure distribution in the lubricant, is calculated. Thus, since all the states of variables are known from dynamic analysis of the mechanical system, the hydrodynamic forces given by (5.19) and (5.20) can be evaluated and introduced as external generalized forces into the system's equations of motion of the multibody mechanical system.

5.4 Transition Between Hydrodynamic and Dry Contact

In this section, a transition model, which combines the squeeze action and the dry contact model, is discussed. This model that considers the existence of lubrication during the free flight trajectory of the journal, prior to contact, and the possibility for dry contact under some conditions, seems to be well fitted to describe revolute joints with clearances in mechanical systems.

In what follows, the expression for the squeeze forces of infinite long journal–bearings is presented. The objective is to evaluate the resulting force from the given state of position and velocity of the journal–bearing. The squeeze forces are exerted when a fluid is squeezed between two approaching surfaces. In journal–bearings, squeeze action is dominant when the relative angular velocity is small compared to the relative radial velocity and, hence, it is reasonable to drop the wedge term in the Reynolds' equation.

In the infinitely long journal–bearing the axial flow is neglected when compared with the circumferential flow, hence the Reynolds' equation reduces to a one-dimensional problem and the pressure field is given by (Hamrock 1994)

$$p = \frac{6\mu R_j^3 \dot{\varepsilon}}{c^2} \frac{\cos\theta(2 - \varepsilon\cos\theta)}{(1 - \varepsilon\cos\theta)^2} \tag{5.21}$$

The resulting force F on the journal that balances the fluid pressure is evaluated as an integral of the pressure field over the surface of the journal:

$$F = \frac{12\pi\mu L R_j^3 \dot{\varepsilon}}{c^2(1 - \varepsilon^2)^{\frac{3}{2}}} \tag{5.22}$$

where μ is the dynamic lubricant viscosity, L is the journal–bearing length, R_j is the journal radius, c is the radial clearance, ε is the eccentricity ratio and $\dot{\varepsilon}$ is the time rate of change of eccentricity ratio.

The direction of the force is along the line of centers that connects the journal and the bearing, which is described by the eccentricity vector. Thus the squeeze force can be introduced into the equations of motion of multibody mechanical systems as a generalized force with the journal and bearing centers as points of action for the force and reaction force, respectively. It should be highlighted that the effect of cavitation is not considered in (5.22), that is, it is assumed that a continuous film exists all around the journal–bearing. However, these conditions are not satisfactory from the physical point of view because the lubricant fluids cannot sustain negative pressures. Based on the mechanics of the journal–bearing, Ravn et al. (2000) included a cavitation effect assuming that negative pressure occurs on the half of journal surface which faces away from the moving direction.

Equation (5.22), which represents the action on the journal that maintains in equilibrium the field pressure, is valid for situations when the load capacity of the wedge effect is negligible when compared to that of the squeeze effect. In the squeeze lubrication, the journal moves along a radial line in the direction of the applied load, thus the film thickness decreases and the fluid is forced to flow up around the journal and out from the ends of the bearing. Since the squeeze force is proportional to the rate of decrease of the fluid film thickness, it is apparent that the lubricant acts as a nonlinear viscous damper resisting the load when the film thickness is decreasing. As the fluid film thickness becomes very thin, that is, the journal is very close to the bearing surface, the force due to the lubricant evaluated from (5.22) becomes very large. Hence a discontinuity appears when the film thickness approaches zero and, consequently, the squeeze force tends to infinity.

For highly loaded contact, the pressure causes elastic deformation of the surfaces, which can be of the same order as the lubricant film thickness. These circumstances are dramatically different from those found in the hydrodynamic regime. The contribution of the theory of elasticity with the hydrodynamic lubrication is called elasto-hydrodynamic lubrication theory (EHL).

Figure 5.5 schematically shows the shape of the lubricant film thickness and the pressure distribution within a typical elasto-hydrodynamic contact. Due to the normal load F the contacting bodies are deformed. The viscous lubricant, adhering to the surfaces of the moving bodies, is dragged into the high-pressure zone of the contact and therefore separates the mating surfaces. The pressure distribution within the elasto-hydrodynamic contact is similar to the dry Hertzian pressure distribution. At the inlet zone, the pressure slowly builds up until it approximately reaches the maximum Hertzian pressure (p). It is, therefore, assumed that for a certain film thickness, called boundary layer, the fluid can no longer be squeezed and the journal and bearing walls are considered to be in contact, in a way similar to that for the dry contact situation presented in Chap. 4. It should be highlighted that the EHL theory is quite complex and its detailed description is beyond the scope of the present work. However, another equivalent model, based on physical reasoning, can be proposed leading to effects similar to those of the EHL theory.

A transition model which combines the squeeze action and the dry contact model is proposed here. Figure 5.6 shows a partial view of a mechanical system representing a revolute clearance joint with lubrication effect, where both the journal and the bearing can have planar motion. The parallel spring–damper element represented

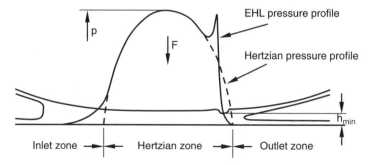

Fig. 5.5 Typical elasto-hydrodynamic contact; qualitative shape of lubricant film and pressure profile

Fig. 5.6 Mechanical system
representing a revolute joint
with lubrication effect

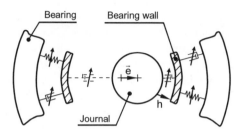

by a continuous line refers to the solid-to-solid contact between the journal and
the bearing wall, whereas the damper represented by a dashed line refers to the lubricated model. If there is no lubricant between the journal and the bearing, the
journal can freely move inside the bearing boundaries. When the gap between the
two elements is filled with a fluid lubricant, a viscous resistance force exists and
opposes the journal motion. Since the radial clearance is specified, the journal and
bearing can work in two different modes. In mode 1, the journal and the bearing
wall are not in contact with each other and they have a relative radial motion.
For the journal–bearing model without lubricant, when $e < c$ the journal is in free
flight motion and the forces associated with the journal–bearing are set to zero. For
lubricated journal–bearing model, the lubricant transmits a force, which must be
evaluated from the state variables of the mechanical system using one of the models
described in the previous section. In mode 2, the journal and the bearing wall are in
contact, thus the contact force between the journal and the bearing is modeled with
the continuous contact force model represented by (3.9).

Since the EHL pressure profile is similar to the Hertzian pressure distribution, it
looks reasonable to change from the squeeze action in the hydrodynamic lubrication
regime to the pure dry contact model. In order to avoid numerical instabilities and
to ensure a smooth transition from pure squeeze model to dry contact model, a
weighted average is used. When the journal reaches the boundary layer, for which
the hydrodynamic theory is no longer valid, the squeeze force model is replaced by
the dry contact force model, as represented in Fig. 5.7a and b. This approach ensures
continuity in the joint reaction forces when the squeeze force model is switched to
dry contact force model. Mathematically the transition force model is expressed by

$$
F = \begin{cases} F_{squeeze} & \text{if} \quad e \leq c \\ \dfrac{(c + e_{t0}) - e}{e_{t0}} F_{squeeze} + \dfrac{e - c}{e_{t0}} F_{dry} & \text{if} \quad c \leq e \leq c + e_{t0} \\ F_{dry} & \text{if} \quad e \geq c + e_{t0} \end{cases} \tag{5.23}
$$

where e_{t0} and e_{t1} are given tolerances for the eccentricity. The values of these parameters must be chosen carefully, since they depend on the clearance size. It should
be noted that the clearance used for the pure squeeze force model is not c but it is
$c + e_{t1}$ instead.

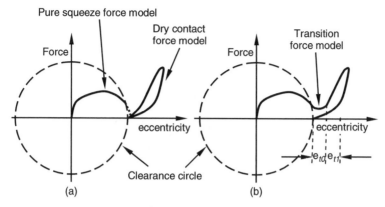

Fig. 5.7 (**a**) Pure squeeze and dry contact force models; (**b**) transition force model between lubricated and dry contact situations

5.5 Kinematic Aspects of the Journal–Bearing Interaction

In order to evaluate the forces produced by the fluid lubricant on the journal–bearing interface, the different dynamic parameters on which these forces depend need to be evaluated. It is clear that the hydrodynamic forces are nonlinear functions of the time parameters, ω, ε, $\dot{\varepsilon}$, γ and $\dot{\gamma}$, which are known at any instant of time from the dynamic system configuration, i.e., these parameters are functions of the system state variables.

Let Fig. 4.4 be redrawn here to highlight the existence of the lubricant fluid in the gap between the journal and the bearing. The two bodies i and j are connected by a lubricated revolute joint, in which the gap between the bearing and the journal is filled with a fluid lubricant. Part of body i is the bearing and part of body j is the journal. The notation used in Fig. 5.8 is the same described in Sect. 4.2.

The parameter ε, which defines the eccentricity ratio, is obtained as the ratio of the distance between the bearing and journal centers by radial clearance, that is,

$$\varepsilon = \frac{e}{c} \tag{5.24}$$

where the eccentricity e is given by (4.3).

The parameter $\dot{\varepsilon}$ can be obtained by differentiating (4.1), and dividing the result by the radial clearance. For the purpose let (4.1) be re-written as

$$\mathbf{e} = \mathbf{r}_j^P + \mathbf{A}_j \mathbf{s}'^P_j - \mathbf{r}_i^P - \mathbf{A}_i \mathbf{s}'^P_i \tag{5.25}$$

then its time derivative is

$$\dot{\mathbf{e}} = \dot{\mathbf{r}}_j^P + \dot{\mathbf{A}}_j \mathbf{s}'^P_j - \dot{\mathbf{r}}_i^P - \dot{\mathbf{A}}_i \mathbf{s}'^P_i \tag{5.26}$$

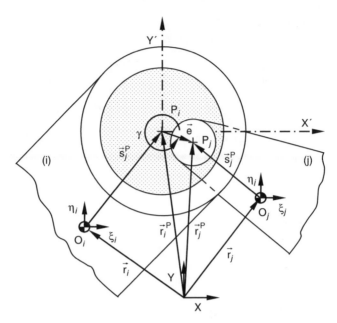

Fig. 5.8 Generic configuration of a dynamically loaded journal–bearing in a multibody mechanical system

Hence the time rate of eccentricity ratio is given by

$$\dot{\varepsilon} = \frac{\dot{e}}{c} \tag{5.27}$$

The line of centers between the bearing and the journal makes an angle γ with X'-axis as shown in Fig. 5.8. Since the unit radial vector **n** has the same direction as the line of centers, the angle γ can be defined as

$$\begin{bmatrix} cos\ \gamma \\ sin\ \gamma \end{bmatrix} = \begin{bmatrix} n_x \\ n_y \end{bmatrix} \tag{5.28}$$

Thus

$$\gamma = tan^{-1}\frac{n_y}{n_x} \tag{5.29}$$

The parameter $\dot{\gamma}$ can be obtained by differentiating (5.29) with respect to time, yielding

$$\dot{\gamma} = \frac{e_x\dot{e}_y - \dot{e}_xe_y}{e^2} \tag{5.30}$$

The hydrodynamic components of force resulting from pressure field projected onto the X and Y directions given by (5.19) and (5.20) act on the journal center.

Fig. 5.9 Vectors of forces
that act on the journal and
bearing

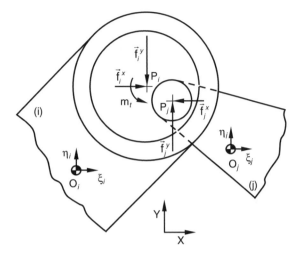

Thus these forces have to be transferred to the centers of mass of the bearing and the journal bodies. From Fig. 5.9, the forces and moments that act on the center of mass of journal body are given by

$$
\begin{bmatrix} f_j^x \\ f_j^y \\ m_j \end{bmatrix} = \begin{bmatrix} F_x \\ F_y \\ -(\xi_j^P \sin\varphi_j + \eta_j^P \cos\varphi_j) F_x + (\xi_j^P \cos\varphi_j - \eta_j^P \sin\varphi_j) F_y \end{bmatrix} \tag{5.31}
$$

and for the center of mass of bearing body, at point O_i,

$$
\begin{bmatrix} f_i^x \\ f_i^y \\ m_i \end{bmatrix} = \begin{bmatrix} -F_x \\ -F_y \\ (\xi_i^P \sin\varphi_i + \eta_i^P \cos\varphi_i + e_y) F_x - (\xi_i^P \cos\varphi_i - \eta_i^P \sin\varphi_i + e_x) F_y \end{bmatrix}
$$
$$\tag{5.32}$$

The transport moment produced by transferring the forces from the center of journal to the center of the bearing is given by $m_t = e_y F_x - e_x F_y$, which is already included in the force vector given by (5.32).

5.6 Application Example 1: Simple Journal–Bearing

In this section, as an application example, a simple journal–bearing subjected to a constant and unidirectional external load is considered, as shown in Fig. 5.10. The journal–bearing under constant unidirectional load contains both the dynamic characteristics within the transient period and the steady hydrodynamic characteristics within the steady-state period.

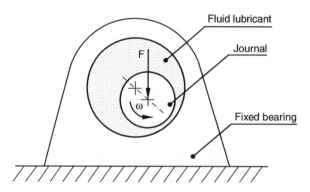

Fig. 5.10 Simple journal–bearing subjected to a constant external load

The journal–bearing properties and initial conditions are listed in Table 5.1. Initially the journal and bearing centers coincide. The oil fluid used in the present example is a SAE 40 multigrade, which is recommended for small combustion engines, and at 40°C the viscosity is equal to 400 cP. In order to analyze the performance of the models proposed, the simple journal–bearing performance is described by the journal center trajectory inside the bearing boundaries, as represented in Fig. 5.11a–e, and by the horizontal and vertical component of the reaction fluid force, shown in Figs. 5.12 and 5.13, respectively.

For the hydrodynamic journal–bearing models that use the Sommerfeld's boundary conditions, the journal center oscillates around its equilibrium position, as observed in Fig. 5.11a and c, whereas for the Gümbel's conditions, after an initial overshoot and transient period, the journal reaches its final equilibrium position, displayed in Fig. 5.11b and d. In the equilibrium position, the squeeze effect becomes null, that is, $\dot{\varepsilon} = 0$. Hence the forces generated by the wedge action balances the external applied load, which in this particular situation is $F_X = 0$, as observed in Fig. 5.12b and d, and $F_Y = F$, as shown in Fig. 5.13b and d. This is expected since the steady-state position is reached and the journal rotates about its whirl. When the Pinkus and Sternlicht hydrodynamic model is used, the journal also reaches the final equilibrium position but with lower damping, as shown in Figs. 5.11e, 5.12e and 5.13e. Indeed this model predicts lower damping and higher fluctuations in the transient phase which seems to be more realistic since in practical cases the oscillation and instability of the journal–bearing are among the main concerns of any analysis. The overall results are corroborated by published works about this field (Alshaer and Lankarani 2001).

Table 5.1 Dynamic properties of the simple journal–bearing

External load	30 N	Journal mass	0.13 kg
Bearing radius	10.0 mm	Journal rotational inertia	$2.5 \times 10^{-4}\, \mathrm{kg}\, m^2$
Journal radius	9.8 mm	Journal angular speed	500 rpm
journal–bearing length	40.0 mm	Oil viscosity at 40°C	400 cP

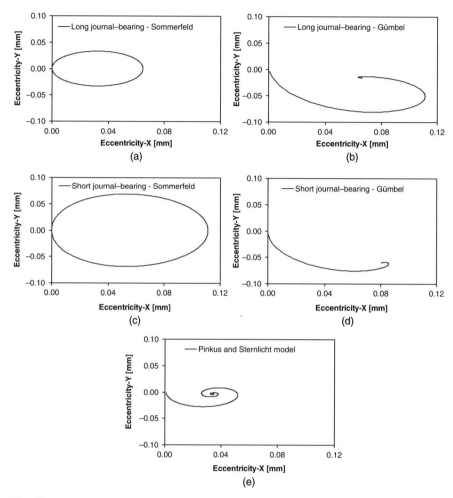

Fig. 5.11 Trajectory of the journal center inside the bearing boundaries for the different dynamically loaded journal–bearing models: (**a**) infinitely long journal–bearing with Sommerfeld's conditions; (**b**) infinitely long journal–bearing with Gümbel's conditions; (**c**) infinitely short journal–bearing with Sommerfeld's conditions; (**d**) infinitely short journal–bearing with Gümbel's conditions; (**e**) Pinkus and Sternlicht hydrodynamic model

When observing the results, it should be pointed out that from the physical point of view a simple journal–bearing subjected to a constant and unidirectional external load corresponds to the free vibration system. The journal is under the influence of a suddenly applied force, which means that the journal is pulled out from the position of stable equilibrium by a small amount and released. A closer look at the forces, displayed in Fig. 5.12, shows that in all models F_X converges or oscillates about $F_X = 0$ N while Fig. 5.13 shows that the trend is convergence or oscillation of F_Y about $F_X = 31.275$ N.

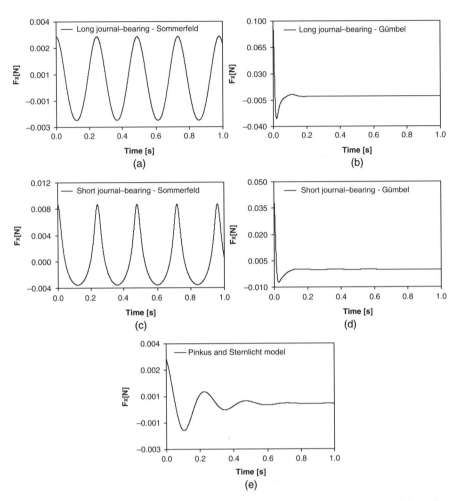

Fig. 5.12 Horizontal reaction force on the journal for the different dynamically loaded journal–bearing models: (**a**) infinitely long journal–bearing with Sommerfeld's conditions; (**b**) infinitely long journal–bearing with Gümbel's conditions; (**c**) infinitely short journal–bearing with Sommerfeld's conditions; (**d**) infinitely short journal–bearing with Gümbel's conditions; (**e**) Pinkus and Sternlicht hydrodynamic model

The journal–bearing performance for the hydrodynamic models with Sommerfeld's boundary conditions corresponds to a free vibration without damping, in which the journal is pulled out of its equilibrium position and then released without initial velocity and, consequently, the motion persists forever. The undamped free vibration, being periodic, is represented by a rotating vector, the end of which describes the circle observed in Fig. 5.11a and c. Yet the Gümbel's solutions, shown in Fig. 5.11b and d, and the Pinkus and Sternlicht model represent a damped free vibration system. In the later model the end point of the rotating vector describes the logarithmic spiral, displayed in Fig. 5.11e. The damped cycle path has its pole at the

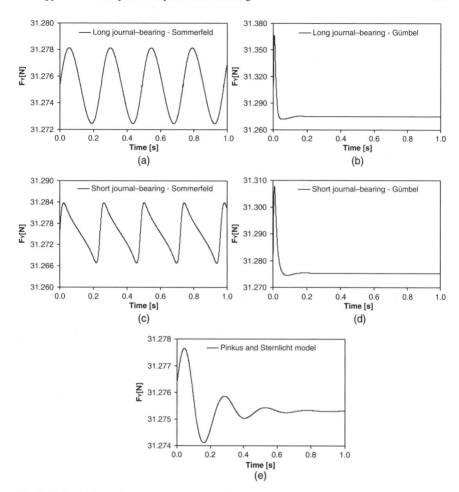

Fig. 5.13 Vertical reaction force on the journal for the dynamically loaded journal–bearing models: (**a**) infinitely long journal–bearing with Sommerfeld's conditions; (**b**) infinitely long journal–bearing with Gümbel's conditions; (**c**) infinitely short journal–bearing with Sommerfeld's conditions; (**d**) infinitely short journal–bearing with Gümbel's conditions; (**e**) Pinkus and Sternlicht hydrodynamic model

steady-state equilibrium position. In fact, the damped oscillations play an important role in the various forms of hydrodynamic instability and vibration, particularly in lightly loaded journal–bearings.

The final equilibrium position clearly depends on the applied load, physical and dynamic properties of the journal–bearing and the hydrodynamic model used. The steady-state equilibrium position does not occur in a dynamically loaded journal–bearing because the applied load varies in both magnitude and direction. This issue is discussed in the next section where a simple planar mechanical system with a dynamically loaded journal–bearing is considered.

5.7 Demonstrative Example 2: Slider–Crank Mechanism

The slider–crank mechanism used in the previous chapter is selected again here as an application example to study the influence of modeling the lubricated revolute joints in the dynamic behavior of mechanical systems. Figure 5.14 shows the configuration of the slider–crank model. The revolute joint between the connecting rod and slider is modeled as lubricated joint. Initially the journal and bearing centers coincide. The crank is the driving link and rotates with a constant angular velocity of 5000 rpm. Furthermore the initial conditions necessary to start the dynamic analysis are obtained from kinematic analysis of the mechanism in which all the joints are considered as ideal joints. The dynamic parameters used in the simulations are listed in Table 5.2.

In order to better understand the influence of the lubricated revolute joint models on the dynamic performance of the slider–crank mechanism, some of the results relative to the dry contact situation discussed in the previous chapter are presented again. Thus, in what follows, four different situations are analyzed and discussed. In the first one, the revolute joint is modeled as dry frictionless contact and the normal contact force law is given by Lankarani and Nikravesh model and expressed by (3.9). In the second case, besides this contact force model, the modified Coulomb friction law given by (3.16) is also included. In the third situation, the revolute clearance joint is modeled by using the transition model, that is, a combination between the dry contact force model and the pure squeeze force model presented in Sect. 5.4. In the fourth simulation the clearance revolute joint is modeled with the hydrodynamic Pinkus and Sternlicht model given by (5.14)–(5.18).

The dynamic behavior of the slider–crank mechanism simulations is quantified by plotting the values of the slider velocity and acceleration and the moment that acts on the crank, represented by Figs. 5.15–5.17. Additionally the journal center trajectories as well as the Poincaré maps are plotted in Figs. 5.18 and 5.19. The results are relative to two complete crank rotations after the transient effect has fadeout.

When observing the slider velocity, slider acceleration and crank moment diagrams, the influence of the joint clearance model is evident. For the cases of dry contact models with and without friction, displayed in Figs. 5.15a, b, 5.16a, b and 5.17a, b, high peaks on the acceleration and moment are visible, which means that

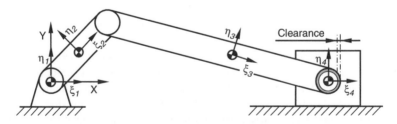

Fig. 5.14 Slider–crank mechanism with a lubricated revolute clearance joint

Table 5.2 Dynamic properties of the simple journal–bearing

Bearing radius	10.0 mm	Young's modulus	207 GPa
Journal radius	9.5 mm	Poisson's ratio	0.3
Journal–bearing length	40.0 mm	Baumgarte - α	5
Restitution coefficient	0.9	Baumgarte - β	5
Friction coefficient	0.03	Integration step size	0.00001 s
Oil viscosity at 40°C	400 cP	Integration tolerance	0.000001 s

the journal and bearing impact each other. Periods of continuous or permanent contact between the journal and bearing, shown in Figs. 5.18a and b, can also be observed. Furthermore the dynamic response of the slider–crank mechanism for dry contact model is clearly nonperiodic, as shown in the corresponding Poincaré maps, presented in Fig. 5.19a and b. However, when the friction effect is included, the peaks on the slider accelerations and, consequently, on the crank moments are reduced, as observed in Figs. 5.16b and 5.17b.

When the transition model is employed, the peaks on the slider acceleration and crank moment curves are smaller when compared to those obtained for dry contact models, as observed in Figs. 5.16c and 5.17c. If the impact forces between the journal and bearing are associated with noise, the fluid clearly acts as a filter in so far as the high frequencies or disturbances are removed or at least reduced. Similar to the case of dry contact with friction simulation, in the transition model the journal

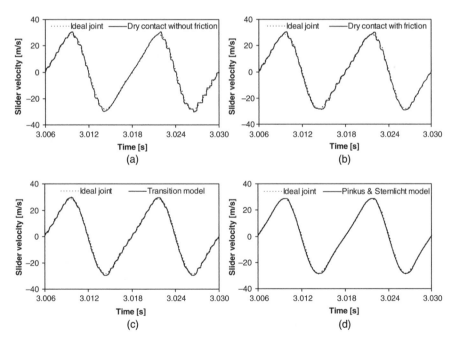

Fig. 5.15 Slider velocity: (**a**) dry contact without friction; (**b**) dry contact with friction; (**c**) transition model; (**d**) Pinkus and Sternlicht hydrodynamic model

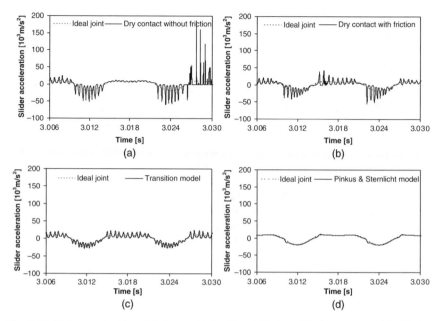

Fig. 5.16 Slider acceleration: (**a**) dry contact without friction; (**b**) dry contact with friction; (**c**) transition model; (**d**) Pinkus and Sternlicht hydrodynamic model

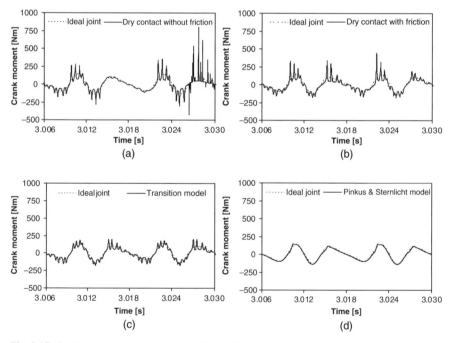

Fig. 5.17 Crank moment: (**a**) dry contact without friction; (**b**) dry contact with friction; (**c**) transition model; (**d**) Pinkus and Sternlicht hydrodynamic model

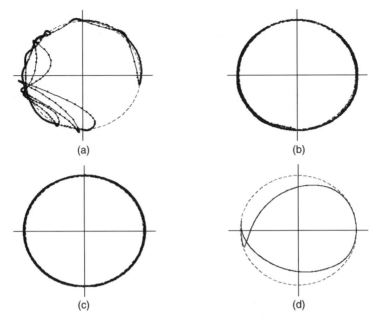

Fig. 5.18 Journal center path: (**a**) dry contact without friction; (**b**) dry contact with friction; (**c**) transition model; (**d**) Pinkus and Sternlicht hydrodynamic model

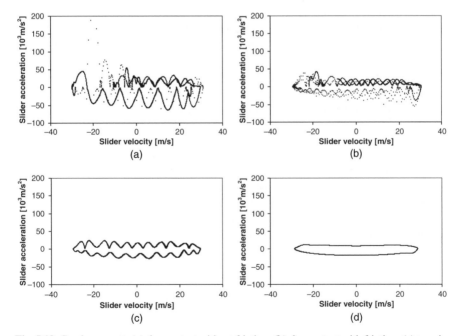

Fig. 5.19 Crank moment: (**a**) dry contact without friction; (**b**) dry contact with friction; (**c**) transition model; (**d**) Pinkus and Sternlicht hydrodynamic model

and bearing are in continuous or permanent contact during all the simulation as is visible in Fig. 5.18c. Hence the system behavior tends to be periodic, as illustrated in the corresponding Poincaré map in Fig. 5.19c.

In the case of the hydrodynamic force model given by Pinkus and Sternlicht, the overall results of the slider–crank mechanism simulation shown in Figs. 5.15d, 5.16d and 5.17d are quite close to those obtained for ideal joints. Clearly, in the case of lubricated revolute joints, the reaction forces developed are much smoother and the peak values are practically removed when compared with the unlubricated case. The system response is clearly periodic or regular, but some numerical instability is observed when the journal moves close to the bearing wall because the system becomes very stiff for a large eccentricity. In the hydrodynamic model, the fluid acts as a nonlinear spring–damper element introducing stiffness and damping to the system. This means that the lubricant absorbs part of the energy produced by the slider–crank mechanism. This is one of the main advantages for the use of fluid lubricants in the joints of machines and mechanisms. In order to better understand the efficiency and accuracy of the lubricated revolute joint using the Pinkus and Sternlicht model, the slider–crank model is considered again for another study. Table 5.3 shows the properties for the new set of study cases.

The performance of the slider–crank mechanism is quantified by plotting the reaction force produced by the fluid on the lubricated joint and the driving crank moment, displayed in Fig. 5.20a and b. In addition, the journal center orbit inside the bearing limits and the minimum oil film thickness are also plotted in Fig. 5.21a and b. The time interval used corresponds to two complete crank rotations. The results are compared to those obtained when the system is modeled with ideal joints only.

From Fig. 5.20a and b it is clear that the results are of the same order of magnitude as those obtained with ideal joints. The first and the second crank cycles show the same results which indicate that the system has reached steady-state operation conditions. This can be confirmed by the orbit of the journal center relative to the bearing center, in which the journal moves far away from the bearing wall, meaning that there is always a minimum film lubricant in between the two bodies, as shown in Fig. 5.21a and b. Since the load on the journal–bearing under consideration is not constant in direction and magnitude, the journal center describes a trajectory within the bearing boundaries as displayed in Fig. 5.21a. This means that the steady-state equilibrium is not reached, which results in a time-dependent loci of the journal center inside the bearing.

The practical criterion for determining whether or not a journal–bearing is operating satisfactorily is the value of the minimum film thickness, which is probably the most important parameter in the performance of journal–bearings.

Table 5.3 Parameters used in the simulations of the slider–crank mechanism

Bearing radius	10.0 mm
Radial clearance	0.2 mm
Journal–bearing length	40.0 mm
Oil viscosity	400 cP

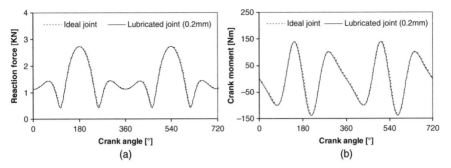

Fig. 5.20 (**a**) Reaction force developed in the lubricated revolute joint, $c = 0.2\,mm$, $\mu = 400\,cP$; (**b**) driving crank moment, $c = 0.2\,mm$, $\mu = 400\,cP$

The minimum film thickness for an aligned journal–bearing is given by

$$h_{min} = c(1 - \varepsilon) \qquad (5.33)$$

where ε is the eccentricity ratio and c is the radial clearance. For safe journal–bearings performance, a minimum film thickness is required.

The safe allowable film thickness depends on the surface finish of the journal. Hamrock (1994) suggests that the safe film thickness should be greater than 2.5 µm. In practical engineering design, it is recommended that the safe film thickness should be at least 0.00015 mm/mm of bearing diameter (Phelan 1970). Similar value for the minimum film thickness can be obtained using the ESDU 84031 Tribology Series (1991) design criterion, shown in Fig. 5.22, which depends on the size of the journal–bearing.

The surface quality of the journal–bearing should be consistent with the expected film thickness and it is, therefore, usual for smaller journal–bearings to be finished to higher standards than larger journal–bearings. ESDU 84031 Tribology Series (1991) suggests that the surface roughness value of the journal–bearings should be not

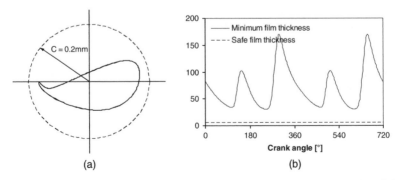

Fig. 5.21 (**a**) Journal center trajectory inside the bearing, $c = 0.2\,mm$, $\mu = 400\,cP$; (**b**) minimum and safe oil film thickness, $c = 0.2\,mm$, $\mu = 400\,cP$

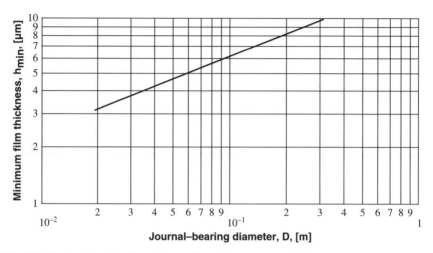

Fig. 5.22 Design chart for the minimum film thickness (adapted from ESDU 84031 Tribology Series 1991)

worse than 1/20 of the minimum film thickness. Thus for the journal–bearing considered here, the safe film thickness that ensures good operating conditions is close to 3 μm. For the present example, the film thickness is always larger than this safe film thickness as demonstrated in Fig. 5.21b.

Equations (5.14)–(5.18) show that the parameters influencing the journal–bearing performance are the oil viscosity μ, the radial clearance c, the bearing length L, the journal radius R_J and the dynamic journal–bearing parameters ω, ε, $\dot{\varepsilon}$, γ and $\dot{\gamma}$. Since the dynamic parameters of the journal–bearing depend directly on the system configuration, the radial clearance size and the oil viscosity are the only possible variables. Thus, in the following, several simulations of the slider–crank mechanism for different values of clearance size and oil viscosity are performed. The behavior of the mechanism is quantified by measuring the values of joint reaction force, driving crank moment and journal center trajectory. The values used for the radial clearance size are 0.5 and 0.1 mm, while the oil viscosity values are 400 and 40 cP.

There are some important differences between the results presented above, namely in what concerns the radial clearance size and oil viscosity influence on the joint reaction force and crank moment; such differences are graphically displayed in Figs. 5.23–5.28. The journal–bearing clearance is an important factor for the satisfactory operation of the journal–bearings. Small values of clearance can give rise to high journal–bearing temperatures while large values of clearance can mean excessive lubricant flow rates. Furthermore the results clearly show the sensitivity of the system response with different values of viscosity. As expected, with low viscosity the journal and the bearing walls are closer than with high viscosity, which suggests the possibility of metal-to-metal contact, especially visible in Figs. 5.26–5.28. Moreover there are some numerical instabilities associated with the lubricated model, namely when the oil viscosity is low, the clearance is too

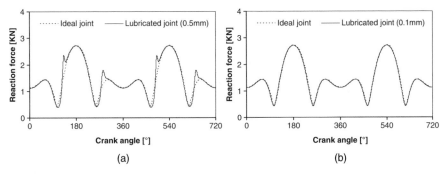

Fig. 5.23 Reaction force developed in the lubricated revolute joint: (**a**) $c = 0.5\,\mathrm{m}m$, $\mu = 400\,\mathrm{c}P$; (**b**) $c = 0.1\,\mathrm{m}m$, $\mu = 400\,\mathrm{c}P$

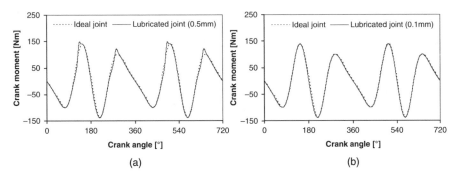

Fig. 5.24 Driving crank moment: (**a**) $c = 0.5\,\mathrm{m}m$, $\mu = 400\,\mathrm{c}P$; (**b**) $c = 0.1\,\mathrm{m}m$, $\mu = 400\,\mathrm{c}P$

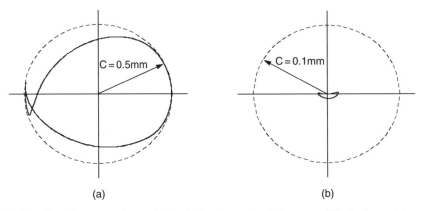

Fig. 5.25 Journal center trajectory inside the bearing: (**a**) $c=0.5\,\mathrm{m}m$, $\mu=400\,\mathrm{c}P$; (**b**) $c=0.1\,\mathrm{m}m$, $\mu=400\,\mathrm{c}P$

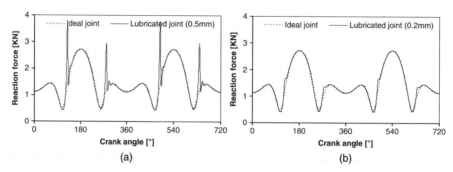

Fig. 5.26 Reaction force developed in the lubricated revolute joint: (**a**) $c = 0.5\,\mathrm{mm}$, $\mu = 40\,\mathrm{c}P$; (**b**) $c = 0.2\,\mathrm{mm}$, $\mu = 40\,\mathrm{c}P$

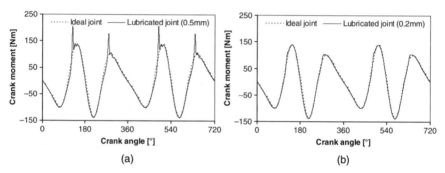

Fig. 5.27 Driving crank moment: (**a**) $c = 0.5\,\mathrm{mm}$, $\mu = 40\,\mathrm{c}P$; (**b**) $c = 0.2\,\mathrm{mm}$, $\mu = 40\,\mathrm{c}P$

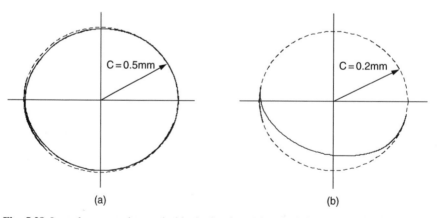

Fig. 5.28 Journal center trajectory inside the bearing: (**a**) $c = 0.5\,\mathrm{mm}$, $\mu = 40\,\mathrm{c}P$; (**b**) $c = 0.2\,\mathrm{mm}$, $\mu = 40\,\mathrm{c}P$

large or for a combination of these two factors. These numerical difficulties are well represented by some peaks in the joint reaction force and crank moment diagrams, because the journal and bearing walls are very close to each other. On the other hand, when the two elements are very close, the hydrodynamic lubrication theory is no longer valid and the EHL theory must be taken into account (Rahnejat 2000).

5.8 Summary

A comprehensive methodology for modeling lubricated revolute joints in multibody mechanical systems was presented in this chapter. A simple journal–bearing and the slider–crank mechanism were chosen as demonstrative examples of the application of the techniques used. Furthermore a parametric study of the slider–crank mechanism for different values of clearance size and oil viscosity was performed in order to provide the results necessary for the comprehensive discussion of the models proposed.

For dynamically loaded journal–bearings, the classic analysis problem consists in predicting the motion of the journal center under arbitrary and known loading. On the contrary, in the present analysis the time-variable parameters are known from the dynamic analysis and the instantaneous force on the journal–bearing is calculated on the bodies of the models proposed. The forces built up by the lubricant fluid are evaluated from the state of variables of the system and included in the equations of motion.

Both squeeze and wedge hydrodynamic effects are included in the analysis of dynamically loaded journal–bearings. It should be mentioned that the methodology presented here uses the superposition principle of wedge effect entrainment and squeeze-film effect to calculate the load capacity. This is only approximate and has been adopted in the work in order to lead to an analytic solution, rather than to a purely numerical one. Furthermore the methodology is only applicable to long bearings, which is seldom the case in most engine bearings, but it can be useful for many other mechanisms. This is the compromise used here for the sake of computational efficiency.

In an application for a slider–crank mechanism, for instance, the reaction moment necessary to drive the crank with a constant angular velocity is of the same order of magnitude as the one in the case of an ideal joint, meaning that the global motion of the slider–crank mechanism with a lubricated joint is periodic or regular. The lubricant acts as a nonlinear spring–damper in so far as lubricated journal–bearing absorbs some of the energy produced by the slider when it accelerates or decelerates, which results in lower reaction moment compared to ideal joints. The lubricant introduces effective stiffness and damping to the slider–crank mechanism. Therefore a hydrodynamic fluid film journal–bearing exhibits a damping effect that plays an important role in the stability of the mechanical components.

The transition model that takes into account the existence of the lubricant during the free flight of the journal and the possibility for dry contact under some conditions seems to be well fitted to describe joints with clearances. The similarity of the forces is notable in the transition model when compared with those of the EHL theory.

The minimum film thickness is a fundamental parameter in the study of a hydrodynamic journal–bearing. The value of the minimum oil film thickness defines the kind of regime of lubrication which is present in the journal–bearing, namely thick-film lubrication, where the journal–bearing surfaces are totally separated by the lubricant and thin-film lubrication in which the field of pressure developed produces elastic deformation. The lubrication between two moving surfaces can shift between these two regimes, depending on the velocity, lubricant viscosity and roughness of surfaces.

The viscosity, which is associated with the resistance to the flow, is one of the most important properties of the fluid lubricants employed in journal–bearings. Actually the behavior of the journal–bearings is quite sensitive to the oil viscosity. As expected, with low working viscosity, the journal and the bearing walls are closer than with high viscosity, which can lead to the possibility of metal-to-metal contact. With higher viscosity there are higher reaction moments produced in systems. This is due to the fact that the lubricant becomes more rigid and absorbs less energy from the system. Moreover the fluid film in conventional journal–bearings can reach a very small thickness and the pressure gradient generated under some conditions can produce elastic deformations of the same order of the magnitude of the film thickness. Thus for a rigorous analysis the elastic deformation must also be considered.

The clearance size also has an important effect on the performance of the journal–bearings. Higher clearance leads to a larger space for the lubricant to flow. The position of the journal loci moves within the bearing, so there is less pressure produced, the motion of the journal being a result of that. Also lower clearance means higher hydrodynamic forces, higher system rigidity and consequently higher reaction forces on the system. Lower clearance produces fewer fluctuations due to the increased damping, which means less vibration and instability problems.

The quantitative solutions for lubricated journal–bearings in mechanical systems presented in this chapter are quite useful in so far as they are given in a closed form. The hydrodynamic model for lubricated revolute joints in mechanical systems is numerically efficient and fast because the pressure distribution does not need to be evaluated. Furthermore the methodology is easy and straightforward to implement in a computational code because resultant forces due to the fluid action are obtained in explicit form. Some numerical difficulties can be observed when either the fluid viscosity is too low or the radial clearance of the journal–bearing is too large, which leads to large eccentricities and, consequently, the system becomes stiff.

References

Alshaer BJ, Lankarani HM (2001) Formulation of dynamic loads generated by lubricated journal bearings. Proceedings of DETC'01, ASME 2001 design engineering technical conferences and computers and information in engineering conference, Pittsburg, PA, September 9–12, 8pp.

Bauchau OA, Rodriguez J (2002) Modelling of joints with clearance in flexible multibody systems. International Journal of Solids and Structures 39:41–63.

Boker JF (1965) Dynamically loaded journal bearings: mobility method of solution. Journal of Basic Engineering 4:537–546.

Claro JCP (1994) Reformulação de método de cálculo de chumaceiras radiais hidrodinâmicas—análise do desempenho considerando condições de alimentação. Ph.D. Dissertation, University of Minho, Guimarães, Portugal.

Costa LAM (2000) Análise do desempenho de chumaceiras radiais hidrodinâmicas considerando efeitos térmicos. Ph.D. Dissertation, University of Minho, Guimarães, Portugal.

Dowson D, Taylor CM (1979) Cavitation in bearings. Annual review of fluid mechanics, edited by JV VanDyke et al., Vol. 11, pp. 35–66. Annual Reviews Inc., Palo Alto, CA.

Dubois GB, Ocvirk FW (1953) Analytical derivation and experimental evaluation of short-bearing approximation for full journal bearings, NACA Rep. 1157.

Elrod HG (1981) A cavitation algorithm. Journal of Lubrication Technology 103:350–354.

ESDU 84031 Tribology Series (1991) Calculation methods for steadily loaded axial groove hydrodynamic journal bearings. Engineering Sciences Data Unit, London, England.

Flores P, Ambrósio J (2004) Revolute joints with clearance in multibody systems. Computers and Structures, Special Issue: Computational Mechanics in Portugal 82(17–18):1359–1369.

Frêne J, Nicolas D, Degneurce B, Berthe D, Godet M (1997) Hydrodynamic lubrication—bearings and thrust bearings. Elsevier, Amsterdam, The Netherlands.

Goenka PK (1984) Analytical curve fits for solution parameters of dynamically loaded journal bearings. Journal of Tribology 106:421–428.

Hamrock BJ (1994) Fundamentals of fluid film lubrication. McGraw-Hill, New York.

Miranda AAS (1983) Oil flow, cavitation and film reformulation in journal bearings including interactive computer-aided design study. Ph.D. Dissertation, Department of Mechanical Engineering, University of Leeds, United Kingdom.

Mistry K, Biswas S, Athre K (1997) A new theoretical model for analysis of the fluid film in the cavitation zone of a journal bearing. Journal of Tribology 119:741–746.

Nikravesh PE (1988) Computer-aided analysis of mechanical systems. Prentice Hall, Englewood Cliffs, NJ.

Phelan RM (1970) Fundamentals of mechanical design, Third edition. McGraw-Hill, New York.

Pinkus O, Sternlicht SA (1961) Theory of hydrodynamic lubrication. McGraw-Hill, New York.

Rahnejat H (2000) Multi-body dynamics: historical evolution and application. Proceedings of the Institution of Mechanical Engineers, Journal of Mechanical Engineering Science 214(C):149–173.

Ravn P (1998) A continuous analysis method for planar multibody systems with joint clearance. Multibody System Dynamics 2:1–24.

Ravn P, Shivaswamy S, Alshaer BJ, Lankarani HM (2000) Joint clearances with lubricated long bearings in multibody mechanical systems. Journal of Mechanical Design 122:484–488.

Sommerfeld A (1904) Zur hydrodynamischen theorie der schmiermittelreibung. Zeitschrift für Angewandte Mathematik und Physik 50:97–155.

Woods CM, Brewe DE (1989) The solution of the Elrod algorithm for a dynamically loaded journal bearing using multigrid techniques. Journal of Tribology 111:302–308.

Chapter 6
Spatial Joints with Clearance: Dry Contact Models

The problem of the dynamic behavior of planar multibody systems with clearance joints was developed in the previous chapters. The utility of the methodologies developed is somewhat restricted because they are not valid for spatial multibody systems such as vehicle models, car suspensions and robotic manipulators, where the system motion is not limited to be planar. In fact, even planar systems may exhibit out-of-plane motion due to misalignments, thus justifying the development of mathematical models to assess the influence of the clearance joints in spatial multibody systems. The main purpose of this chapter is to present effective methodologies for spatial multibody systems including both the spherical and the revolute joints with clearance. Due to their relevance for this chapter, some aspects of the multibody formulation for spatial systems, based on the Cartesian coordinates, are reviewed here to introduce the basic aspects on the dynamic modeling of spatial multibody systems with clearance joints (Nikravesh 1988). A brief description of the ideal, or perfect, spherical and spatial revolute joints is presented. Similar to the case of planar formulation, the bodies that constitute the spatial clearance joints are modeled as colliding bodies and contact-impact forces control the dynamic behavior of the joint elements. For this purpose, the joint elements are considered as elastic bodies in contact, in which relative penetration exists, but without deformation. The normal contact force that depends on this pseudo-penetration follows the contact-impact force model proposed by Lankarani and Nikravesh (1990). This force model, which is a function of the bodies' relative motion and of the internal geometry of the joint, leads to the contact forces that are introduced in the system's equations of motion. In this methodology, the clearance plays a key role in the joint kinematics. Simple spatial mechanical systems that describe spatial motion, such as the four-bar mechanism and the double pendulum, are used to illustrate the methodologies and assumptions adopted. In addition, a slider–crank mechanism that describes a planar motion is also considered, as an application example, in order to compare both the planar and spatial formulation for clearance joints.

6.1 Spatial Multibody Systems

This section presents the formulation of the general equations of motion to the spatial dynamic analysis of multibody systems. A simple and brief description of the standard mechanical joints of spatial multibody mechanical systems is presented, namely of the ideal spherical and revolute joints, to emphasize the differences with respect to joints with clearance, introduced later. The methodology presented can be implemented in any general-purpose multibody code, tested in particular in the computer program DAP-3D, which has been developed for the spatial dynamic analysis of general multibody systems (Nikravesh 1988). Due to its simplicity and computational easiness, Cartesian coordinates and Newton–Euler's method are used to formulate the equations of motion of the spatial multibody systems.

Let Fig. 6.1 represent a rigid body i to which a body-fixed coordinate system $(\xi\eta\zeta)_i$ is attached at its center of mass. When Cartesian coordinates are used, the position and orientation of the rigid body must be defined by a set of translational and rotational coordinates. The position of the body with respect to global coordinate system XYZ is defined by the coordinate vector $\mathbf{r}_i = [x \ y \ z]_i^T$ that represents the location of the local reference frame $(\xi\eta\zeta)_i$. The orientation of the body is described by the rotational coordinate's vector $\mathbf{p}_i = [e_0 \ e_1 \ e_2 \ e_3]_i^T$, which is made with the Euler parameters for the rigid body (Nikravesh 1988). Therefore the vector of coordinates that completely describes the rigid body i is

$$\mathbf{q}_i = [\mathbf{r}_i^T \ \mathbf{p}_i^T]_i^T \tag{6.1}$$

A spatial multibody system with nb bodies is described by a set of coordinates $\mathbf{q}$ in the form

$$\mathbf{q} = [\mathbf{q}_1^T, \mathbf{q}_2^T, \ldots, \mathbf{q}_{nb}^T]^T \tag{6.2}$$

The location of point P on body i can be defined by the position vector $\mathbf{s}_i^P$ with respect to the body-fixed reference frame $(\xi\eta\zeta)_i$ and by the global position vector $\mathbf{r}_i$, that is,

$$\mathbf{r}_i^P = \mathbf{r}_i + \mathbf{s}_i^P = \mathbf{r}_i + \mathbf{A}_i \mathbf{s}_i'^P \tag{6.3}$$

Fig. 6.1 Definition of the Cartesian coordinates for a rigid body

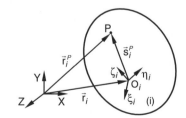

where $\mathbf{A}_i$ is the transformation matrix for body i that defines the orientation of the referential $(\xi\eta\zeta)_i$ with respect to the referential frame XYZ. The transformation matrix is expressed as function of the four Euler parameters as (Nikravesh 1988)

$$\mathbf{A}_i = \begin{bmatrix} e_0^2 + e_1^2 - \frac{1}{2} & e_1 e_2 - e_0 e_3 & e_1 e_3 + e_0 e_2 \\ e_1 e_2 + e_0 e_3 & e_0^2 + e_2^2 - \frac{1}{2} & e_2 e_3 - e_0 e_1 \\ e_1 e_3 - e_0 e_2 & e_2 e_3 + e_0 e_1 & e_0^2 + e_3^2 - \frac{1}{2} \end{bmatrix}_i \tag{6.4}$$

Notice that the vector $\mathbf{s}_i^P$ is expressed in global coordinates whereas the vector $\mathbf{s}_i'^P$ is defined in the body i fixed coordinate system. Throughout the formulation presented in this work, the quantities with $(.)'$ mean that $(.)$ is expressed in local system coordinates.

The velocities and accelerations of body i use the angular velocities $\boldsymbol{\omega}'_i$ and accelerations $\dot{\boldsymbol{\omega}}'_i$ instead of the time derivatives of the Euler parameters, which simplifies the mathematical formulation and does not require the use of mathematical constraint for Euler parameters. The relation between the Euler parameters $\dot{e}_0 + \dot{e}_1 + \dot{e}_2 + \dot{e}_3 = 0$ is implied in the angular velocity and, therefore, is not used explicitly (Nikravesh and Chung 1982). When Euler parameters are employed as rotational coordinates, the relation between their time derivatives, $\dot{\mathbf{p}}_i$, and the angular velocities is expressed by (Nikravesh 1988)

$$\dot{\mathbf{p}}_i = \tfrac{1}{2}\mathbf{L}^T \boldsymbol{\omega}'_i \tag{6.5}$$

where the auxiliary 3×4 matrix $\mathbf{L}$ is a function of Euler parameters (Nikravesh 1988)

$$\mathbf{L}_i = \begin{bmatrix} -e_1 & e_0 & e_3 & -e_2 \\ -e_2 & -e_3 & e_0 & e_1 \\ -e_3 & e_2 & -e_1 & e_0 \end{bmatrix}_i \tag{6.6}$$

The velocities and accelerations of body i are given by vectors

$$\dot{\mathbf{q}}_i = [\dot{\mathbf{r}}_i^T \ \boldsymbol{\omega}'^T_i]_i^T \tag{6.7}$$

$$\ddot{\mathbf{q}}_i = [\ddot{\mathbf{r}}_i^T \ \dot{\boldsymbol{\omega}}'^T_i]_i^T \tag{6.8}$$

In terms of the Cartesian coordinates, the equations of motion of an unconstrained multibody mechanical system are written as

$$\mathbf{M}\ddot{\mathbf{q}} = \mathbf{g} \tag{6.9}$$

where $\mathbf{M}$ is the global mass matrix, containing the mass and moment of inertia of all bodies and $\mathbf{g}$ is a force vector that contains the external and Coriolis forces acting on the bodies of the system.

For a constrained multibody system, the kinematical joints are described by a set of holonomic algebraic constraints denoted as

$$\Phi(\mathbf{q}, t) = \mathbf{0} \tag{6.10}$$

Using the Lagrange multipliers technique the constraints are added to the equations of motion. These are written together with the second time derivative of the constraint equations. Thus the set of equations that describe the motion of the multibody system is

$$\begin{bmatrix} \mathbf{M} & \Phi_\mathbf{q}^T \\ \Phi_\mathbf{q} & 0 \end{bmatrix} \begin{Bmatrix} \ddot{\mathbf{q}} \\ \lambda \end{Bmatrix} = \begin{Bmatrix} \mathbf{g} \\ \gamma \end{Bmatrix} \tag{6.11}$$

where λ is the vector of Lagrange multipliers and γ is the vector that groups all the terms of the acceleration constraint equations that depend on the velocities only, that is,

$$\gamma = -(\Phi_\mathbf{q} \dot{\mathbf{q}})_\mathbf{q} \dot{\mathbf{q}} - \Phi_{tt} - 2\Phi_{\mathbf{q}t} \dot{\mathbf{q}} \tag{6.12}$$

The Lagrange multipliers, associated with the kinematic constraints, are physically related to the reaction forces and moments generated between the bodies interconnected by kinematic joints.

Equation (6.11) is a differential algebraic equation that has to be solved, the resulting accelerations being integrated in time. However, in order to avoid constraints violation during numerical integration, the Baumgarte (1972) stabilization technique is used, (6.11) being modified to

$$\begin{bmatrix} \mathbf{M} & \Phi_\mathbf{q}^T \\ \Phi_\mathbf{q} & 0 \end{bmatrix} \begin{Bmatrix} \ddot{\mathbf{q}} \\ \lambda \end{Bmatrix} = \begin{Bmatrix} \mathbf{g} \\ \gamma - 2\alpha\dot{\Phi} - \beta^2\Phi \end{Bmatrix} \tag{6.13}$$

where α and β are positive constants that represent the feedback control parameters for the velocity and position constraint violations (Baumgarte 1972, Nikravesh 1988). This issue is presented and discussed in detail in Chap. 2, the conclusions being valid for spatial systems also. The same applies to the use of the vector calculus and the numerical methods for planar dynamics of multibody systems presented in previous chapters, which are still adequate for the treatment of spatial systems.

According to the formulation outlined, the dynamic response of multibody systems involves the evaluation of the Jacobian matrix $\Phi_\mathbf{q}$ and vectors $\mathbf{g}$ and γ, in each time step. The solution of (6.13) is obtained for the system accelerations $\ddot{\mathbf{q}}$. These accelerations, together with the velocities $\dot{\mathbf{q}}^*$, are integrated to obtain the new velocities $\dot{\mathbf{q}}$ and positions $\mathbf{q}^*$ for the next time step. This process is repeated until the complete description of system motion is obtained for a selected time interval. Note that, in vector $\dot{\mathbf{q}}^*$, the angular velocities are substituted by the time derivatives of the Euler parameters using (6.5).

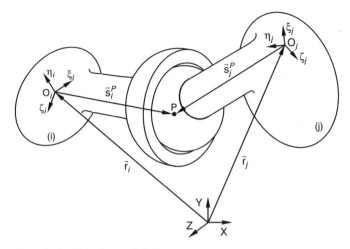

Fig. 6.2 Perfect spherical joint in a multibody system

An ideal or perfect spherical joint, also known as the ball and socket joint, illustrated in Fig. 6.2, constrains the relative translations between two adjacent bodies i and j, allowing only three relative rotations. Therefore the center of the spherical joint, point P, has constant coordinate with respect to any of the local coordinate systems of the connected bodies, i.e., a spherical joint is defined by the condition that the point P_i on body i coincides with the point P_j on body j. This condition is simply the spherical constraint, which can be written in a scalar form as (Nikravesh 1988)

$$\boldsymbol{\Phi}^{(s,3)} \equiv \mathbf{r}_i + \mathbf{A}_i \mathbf{s}_i'^P - \mathbf{r}_j - \mathbf{A}_j \mathbf{s}_j'^P = \mathbf{0} \qquad (6.14)$$

The three scalar constraint equations implied by (6.14) restrict the relative position of points P_i and P_j. Therefore three relative degrees of freedom are maintained between two bodies that are connected by a perfect spherical joint.

An ideal three-dimensional revolute or rotational joint between bodies i and j, shown in Fig. 6.3, is built with a journal–bearing that allows a relative rotation about a common axis, but precludes relative translation along this axis. Equation (6.14) is imposed on an arbitrary point P on the joint axis. Two other points Q_i on body i and Q_j on body j are also arbitrarily chosen on the joint axis. It is clear that vectors $\mathbf{s}_i$ and $\mathbf{s}_j$ must remain parallel. Therefore there are five constraint equations for an ideal three-dimensional revolute joint (Nikravesh 1988):

$$\boldsymbol{\Phi}^{(r,5)} \equiv \begin{cases} \mathbf{r}_i + \mathbf{A}_i \mathbf{s}_i'^P - \mathbf{r}_j - \mathbf{A}_j \mathbf{s}_j'^P = \mathbf{0} \\ \tilde{\mathbf{s}}_i \mathbf{s}_j = \mathbf{0} \end{cases} \qquad (6.15)$$

Note that the cross product in (6.15) has only two independent constraints, the third being linearly dependent on the first two. The five scalar constraint equations

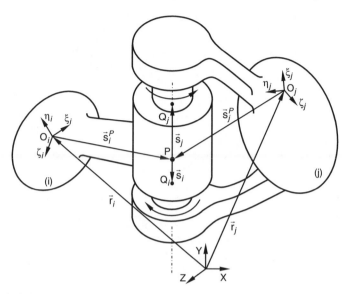

Fig. 6.3 Ideal three-dimensional or spatial revolute joint in a multibody system

yield only one relative degree of freedom for this joint, that is, rotation about the common axis of the revolute joint.

6.2 Spherical Joint with Clearance

In this section, a mathematical model of spherical joint with clearance for spatial multibody systems is presented. In standard multibody models, it is assumed that the connecting points of two bodies, linked by an ideal or perfect spherical joint, are coincident. The introduction of the clearance in a spherical joint separates these two points and the bodies become free to move relative to each other. Hence the three kinematic constraints shown in (6.14) are removed and three relative DOF are allowed instead. A spherical joint with clearance does not constrain any DOF from the system like the ideal spherical joint. In a spherical clearance joint, the dynamics of the joint is controlled by contact-impact forces that result from the collision between the bodies connected. Thus these types of joints can be called as force-joints, since they deal with force constraints instead of kinematic constraints.

Figure 6.4 depicts two bodies i and j connected by a spherical joint with clearance. A spherical part of body j, the ball, is inside of a spherical part of body i, the socket. The radii of socket and ball are R_i and R_j, respectively. The difference in radius between the socket and the ball defines the size of radial clearance, $c = R_i - R_j$. The centers of mass of bodies i and j are O_i and O_j, respectively. Body-fixed coordinate systems $\xi\eta\zeta$ are attached at their center of mass, while XYZ represents the global coordinate system. Point P_i indicates the center of the socket, the center of the ball being denoted by P_j. The vector that connects the point P_i to

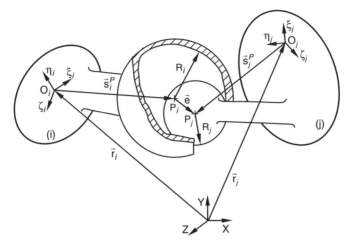

Fig. 6.4 Generic spherical joint with clearance in a multibody system

point P_j is defined as the eccentricity vector, which is represented in Fig. 6.4. Note that, in real mechanisms, the magnitude of the eccentricity is typically much smaller than the radius of the socket and ball.

Similar to the two-dimensional revolute joint, when some amount of clearance is included in a spherical joint, the ball and socket can move relative to each other. Figure 6.5 illustrates the different types of ball motion inside the socket, namely, contact or following mode, free flight mode and impact mode.

In the contact or following mode, the ball and the socket are in permanent contact and a sliding motion relative to each other exists. This mode ends when the ball

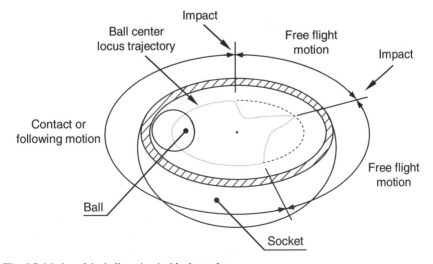

Fig. 6.5 Modes of the ball motion inside the socket

and socket separate from each other and the ball enters the free flight mode. In the free flight motion, the ball moves freely inside the socket boundaries, that is, the ball and the socket are not in contact, hence there is no joint reaction force. In the impact mode, which occurs at the termination of the free flight motion, impact forces are applied to the system. This mode is characterized by an abrupt discontinuity in the kinematic and dynamic responses, and a significant exchange of momentum between the two impacting bodies is observed. At the termination of the impact mode, the ball can enter either the free flight or the following mode.

In what follows, some of the most relevant kinematic aspects related to the spherical clearance joint are presented. As displayed in Fig. 6.4, the eccentricity vector $\mathbf{e}$, which connects the centers of the socket and the ball, is given by

$$\mathbf{e} = \mathbf{r}_j^P - \mathbf{r}_i^P \tag{6.16}$$

where both $\mathbf{r}_j^P$ and $\mathbf{r}_i^P$ are described in global coordinates with respect to the inertial reference frame (Nikravesh 1988):

$$\mathbf{r}_k^P = \mathbf{r}_k + \mathbf{A}_k \mathbf{s}_k'^P, \quad (k = i, j) \tag{6.17}$$

The magnitude of the eccentricity vector is evaluated as

$$e = \sqrt{\mathbf{e}^T \mathbf{e}} \tag{6.18}$$

The magnitude of the eccentricity vector expressed in the global coordinates is written as

$$e = \sqrt{(x_j^P - x_i^P)^2 + (y_j^P - y_i^P)^2 + (z_j^P - z_i^P)^2} \tag{6.19}$$

and the time rate of change of the eccentricity in the radial direction, that is, in the direction of the line of centers of the socket and the ball is

$$\dot{e} = \frac{(x_j^P - x_i^P)(\dot{x}_j^P - \dot{x}_i^P) + (y_j^P - y_i^P)(\dot{y}_j^P - \dot{y}_i^P) + (z_j^P - z_i^P)(\dot{z}_j^P - \dot{z}_i^P)}{e} \tag{6.20}$$

in which the dot denotes the derivative with respect to time.

A unit vector $\mathbf{n}$ normal to the collision surface between the socket and the ball is aligned with the eccentricity vector, as observed in Fig. 6.6. Thus

$$\mathbf{n} = \frac{\mathbf{e}}{e} \tag{6.21}$$

Figure 6.6 illustrates the situation in which the socket and the ball bodies are in contact, which is identified by the existence of a relative penetration. The contact or control points on bodies i and j are Q_i and Q_j, respectively. The global position of the contact points in the socket and ball are given by

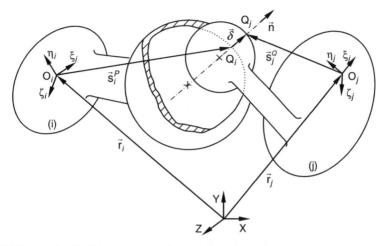

Fig. 6.6 Penetration depth between the socket and the ball during the contact

$$\mathbf{r}_k^Q = \mathbf{r}_k + \mathbf{A}_k \mathbf{s}_k'^Q + R_k \mathbf{n}, \quad (k = i, j) \tag{6.22}$$

where R_i and R_j are radius of the socket and the ball, respectively.

The velocity of the contact points Q_i and Q_j in the global system is obtained by differentiating (6.22) with respect to time, yielding

$$\dot{\mathbf{r}}_k^Q = \dot{\mathbf{r}}_k + \dot{\mathbf{A}}_k \mathbf{s}_k'^Q + R_k \dot{\mathbf{n}}, \ (k = i, j) \tag{6.23}$$

Let the components of the relative velocity of contact points in the normal and tangential direction to the surface of collision be represented by $\mathbf{v}_N$ and $\mathbf{v}_T$, respectively. The relative normal velocity determines whether the contact bodies are approaching or separating, and the relative tangential velocity determines whether the contact bodies are sliding or sticking. The relative scalar velocities, normal and tangential to the surface of collision, are obtained by projecting the relative impact velocity onto the tangential and normal directions, yielding

$$\mathbf{v}_N = [(\dot{\mathbf{r}}_j^Q - \dot{\mathbf{r}}_i^Q)^T \mathbf{n}]\mathbf{n} \tag{6.24}$$

$$\mathbf{v}_T = (\dot{\mathbf{r}}_j^Q - \dot{\mathbf{r}}_i^Q)^T - \mathbf{v}_N \equiv v_T \mathbf{t} \tag{6.25}$$

where $\mathbf{t}$ represents the tangential direction to the impacted surfaces.

From Fig. 6.6 it is clear that the geometric condition for contact between the socket and the ball can be defined as

$$\delta = e - c \tag{6.26}$$

where e is the magnitude of the eccentricity vector given by (6.18) and c is the radial clearance size. It should be noted here that the clearance is taken as a specified parameter. When the magnitude of the eccentricity vector is smaller than the radial clearance size there is no contact between the socket and the ball and, consequently, they can freely move relative to each other. When the magnitude of eccentricity is larger than radial clearance, there is contact between the socket and the ball, the relative penetration being given by (6.26). Then a constitutive contact law, such as the continuous contact force model proposed by Lankarani and Nikravesh (1990), expressed by (3.9), is applied in order to evaluate the contact force developed in the direction perpendicular to the plane of collision. Thus the magnitude of the contact force can be summarized as follows:

$$
F_N =
\begin{cases}
0 & \text{if } \delta < 0 \\
K\delta^n \left[1 + \dfrac{3(1 - c_e^2)}{4} \dfrac{\dot{\delta}}{\dot{\delta}^{(-)}} \right] & \text{if } \delta > 0
\end{cases}
\tag{6.27}
$$

where the generalized parameter K is evaluated by (3.3), c_e is the restitution coefficient, $\dot{\delta}$ is the relative penetration velocity and $\dot{\delta}^{(-)}$ is the initial impact velocity. The nature of the contact forces for spatial systems is identical to those for planar systems, which are presented and discussed in Chap. 3.

The normal and tangential forces at the contact points are represented by $\mathbf{f}_N$ and $\mathbf{f}_T$, respectively. Since these forces do not act through the center of mass of bodies i and j, the moment components for each body need to be evaluated. Furthermore the contribution of the contact forces to the generalized force vector is obtained by projecting the normal and tangential forces onto the X, Y and Z directions. Based on Fig. 6.7, the equivalent forces and moments applied on the center of mass of body i are given by

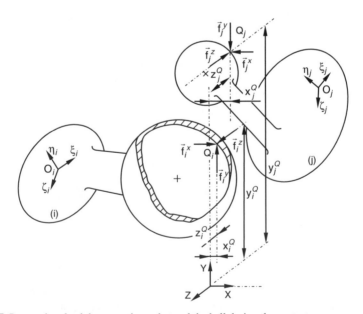

Fig. 6.7 Penetration depth between the socket and the ball during the contact

$$\mathbf{f}_i = \mathbf{f}_N + \mathbf{f}_T \tag{6.28}$$

$$\mathbf{m}_i = \tilde{\mathbf{s}}_i^Q \mathbf{f}_i \tag{6.29}$$

The forces and moments acting on body j are written as

$$\mathbf{f}_j = -\mathbf{f}_i \tag{6.30}$$

$$\mathbf{m}_j = -\tilde{\mathbf{s}}_j^Q \mathbf{f}_i \tag{6.31}$$

For notation purpose the tilde ($\sim$) placed over a vector indicates that the components of the vector are used to generate a skew-symmetric matrix (Nikravesh 1988).

6.3 Spatial Revolute Joint with Clearance

The typical configuration of a spatial revolute joint with clearance is schematically illustrated in Fig. 6.8. The pair of elements in a spatial revolute clearance joint are a cylindrical hole, the bearing, and a cylindrical pin, the journal. The clearance, in a realistic connection, is much smaller than the length of the two cylinders or the nominal radius of the joint elements.

Similar to the spherical clearance joint model development, the two mechanical bodies connected by the joint are modeled as colliding bodies, and, consequently, contact-impact forces control the dynamics of the joint. In the methodology presented here, the contact force model with hysteric damping is used to evaluate the normal contact forces resulting from the interpenetration between the journal and the bearing. For this purpose, the mechanical elements are considered as two rigid bodies in contact that penetrate into each other, without deforming. The normal contact force depends on this pseudo-penetration, according to the model proposed by Lankarani and Nikravesh (1990). Thus it is clear that the spatial revolute joint with clearance does not impose any kinematic constraint to the system, but imposes some force restrictions, limiting the journal movement within the bearing limits.

The model for the spatial revolute clearance joint is more complex than the spherical joint with clearance, because there are more paths of motion for the journal when clearance is present. Four different types of journal motion inside the bearing are considered in the present work, namely: (1) free flight motion where there is

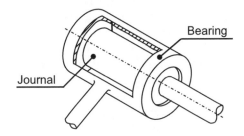

Fig. 6.8 Typical spatial revolute joint with clearance

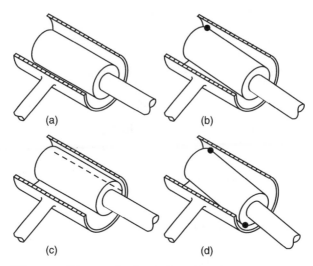

Fig. 6.9 Four different possible scenarios for the journal motion relative to the bearing: (**a**) no contact; (**b**) one point contact; (**c**) line contact; (**d**) two contact points in opposite sides

no contact between the two elements; (2) the journal contacts with the bearing wall at a point; (3) the journal and bearing contact with each other at a line; (4) two contact points between the journal and the bearing wall, but in opposite sides. These four possibilities are illustrated in Fig. 6.9. The dynamic response of the joint is a function of these four scenarios which depend on the system configuration.

In a noncontact situation, no forces are introduced into the system, because the journal moves freely inside the bearing boundaries until it reaches the bearing wall. When the journal and the bearing are in contact with each other, local deformations take place at the contact area and, consequently, contact-impact forces characterize the interaction between the bodies. By evaluating the variation of the contact forces during the contact period, the system response is obtained simply by adding the contact-impact forces to the multibody system equations of motion as external generalized forces. This approach provides accurate results, in so far as the equations of motion are integrated over the period of contact. It, thus, accounts for the changes in the configuration and velocities of the system during that contact.

In order for the spatial revolute clearance joints to be used in the multibody system formulation, it is required that a mathematical model be developed. Figure 6.10 shows a representation of a spatial revolute joint with clearance that connects bodies i and j. The bearing is part of body i and the journal is part of body j. The difference in radius between the bearing and the journal, $c = R_i - R_j$, defines the size of the radial clearance. The center of mass of bodies i and j are O_i and O_j, respectively. Body-fixed coordinate systems $\xi\eta\zeta$ are attached at the center of mass of each body, whereas the XYZ denotes the global coordinate system. The geometric center of the bearing is located at point P which, together with point Q, defines the joint/bearing axis, while points V and W on body j define the journal axis. These points are

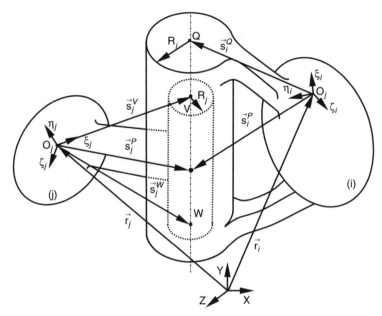

Fig. 6.10 General configuration of a three-dimensional or spatial revolute joint with clearance in a multibody system

located at the top and bottom of journal bases, so that the distance between points V and W defines the length of the joint.

Figure 6.11 shows two different scenarios for the contact between the journal and the bearing. For simplicity, in this figure only the journal and the bearing are

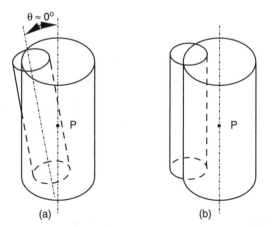

Fig. 6.11 Two different scenarios for contact between the journal and the bearing wall: (**a**) only one journal base (top) contacts with the bearing wall; (**b**) both bases (top and bottom) contact with the bearing wall

represented. In the present methodology, only the top and the bottom journal bases are considered for contact detection between the journal and bearing elements. Furthermore it is assumed that the clearance is much smaller than the dimensions of the bodies, so that the angle between the bearing and journal axes, represented by θ in Fig. 6.11a, is very small and, consequently, both top and bottom journal bases can be assumed to be parallel and perpendicular to the joint axis, as illustrated in Fig. 6.12.

Assuming a local coordinate system $(\xi\eta\zeta)_r$ associated with the revolute clearance joint axis located at point P, the unit coordinate vectors defined along the local axes are $\mathbf{u}'_\xi$, $\mathbf{u}'_\eta$ and $\mathbf{u}'_\xi$, as illustrated in Fig. 6.13. The unit coordinate vector along the ζ_r-axis, $\mathbf{u}'_\xi$, is evaluated as

$$\mathbf{u}'_\xi = \frac{\mathbf{s}'^Q_i - \mathbf{s}'^P_i}{\left\| \mathbf{s}'^Q_i - \mathbf{s}'^P_i \right\|} \tag{6.32}$$

where both vectors $\mathbf{s}'^P_i$ and $\mathbf{s}'^Q_i$ are described in the local coordinate system of body i. The remaining two unit vectors are evaluated according to

$$\begin{cases} \mathbf{u}'_\xi = \mathbf{u}'_\xi \\ \mathbf{u}'_\eta = \mathbf{u}'_\eta \end{cases} \quad \text{if} \quad \mathbf{u}'_\xi = \mathbf{u}'_\zeta \tag{6.33}$$

or

$$\begin{cases} \mathbf{u}'_\xi = \tilde{\mathbf{u}}'_{\zeta_r} \, \mathbf{u}'_{\zeta_i} \\ \mathbf{u}'_\eta = \tilde{\mathbf{u}}'_{\zeta_r} \, \mathbf{u}'_{\zeta_r} \end{cases} \quad \text{if} \quad \mathbf{u}'_\xi \neq \mathbf{u}'_\zeta \tag{6.34}$$

Fig. 6.12 Front and top views for contact between the journal and the bearing: (**a**) only the journal top base contacts with the bearing wall; (**b**) both journal bases (top and bottom) contact with the bearing wall

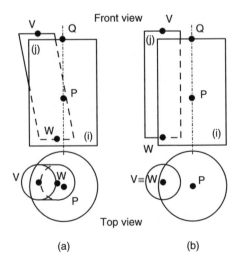

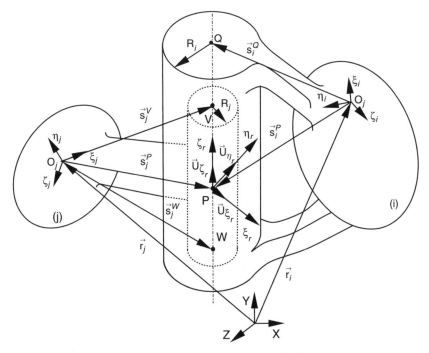

Fig. 6.13 Definition of the local coordinate system associated with the revolute clearance joint and the respective unit vectors

The transformation matrix $\mathbf{A}_{ri}$ from local components $(\xi\eta\zeta)_r$ into the local coordinate system $(\xi\eta\zeta)_i$ is expressed as (Nikravesh 1988)

$$\mathbf{A}_{ri} = [\mathbf{u}'_\xi \ \mathbf{u}'_\eta \ \mathbf{u}'_\xi]^{\mathrm{T}} \tag{6.35}$$

Note that this transformation matrix is constant. Furthermore if $\mathbf{u}'_\xi = \mathbf{u}'_\xi$ the transformation matrix $\mathbf{A}_{ri}$ is the identity matrix $\mathbf{I}$. Thus the matrix that transforms the local vectors $(\xi\eta\zeta)_r$ into the global reference system XYZ is given by

$$\mathbf{A}_r = \mathbf{A}_i \mathbf{A}_{ri} \tag{6.36}$$

The global position of the origin of the local coordinate system $(\xi\eta\zeta)_r$ is

$$\mathbf{r}_i^P = \mathbf{r}_i + \mathbf{A}_i \mathbf{s}'_{i^P} \tag{6.37}$$

In order to define the relative position between the journal and the bearing, it is necessary to express the vectors $\mathbf{s}_j^V$ and $\mathbf{s}_j^W$ in the local coordinate system associated with the joint $(\xi\eta\zeta)_r$. From Fig. 6.13, the global coordinates of points V and W with respect to the inertial reference frame are expressed as

$$\mathbf{r}_j^V = \mathbf{r}_j + \mathbf{A}_j \mathbf{s'}_j^V \tag{6.38}$$

$$\mathbf{r}_j^W = \mathbf{r}_j + \mathbf{A}_j \mathbf{s'}_j^W \tag{6.39}$$

Thus vectors $\mathbf{s}_j^V$ and $\mathbf{s}_j^W$ expressed in the global coordinate system are

$$\mathbf{s}_r^V = \mathbf{r}_j^V - \mathbf{r}_i^P \tag{6.40}$$

$$\mathbf{s}_r^W = \mathbf{r}_j^W - \mathbf{r}_i^P \tag{6.41}$$

When expressed in the local coordinate system of the joint, these vectors are given by

$$\mathbf{s'}_r^V = \mathbf{A}_r^T \mathbf{s}_r^V \tag{6.42}$$

$$\mathbf{s'}_r^W = \mathbf{A}_r^T \mathbf{s}_r^W \tag{6.43}$$

The vectors given by (6.42) and (6.43) define the coordinates of points V and W of the journal, expressed in terms of the local coordinate system associated with the joint, that is, $(\xi\eta\zeta)_r$.

Figure 6.14 depicts a configuration for the system in which both top and bottom journal bases contact with the bearing wall. The eccentricity vectors at the top and bottom journal bases, $\mathbf{e}_r^V$ and $\mathbf{e}_r^W$, are given by the projection of the vectors $\mathbf{s}_r^V$ and $\mathbf{s}_r^W$ onto the local axes ξ_r and η_r, yielding

$$\mathbf{e}_r^V = \{(\mathbf{s'}_r^V)_{\xi_r} \quad (\mathbf{s'}_r^V)_{\eta_r} \quad 0\}^T \tag{6.44}$$

$$\mathbf{e}_r^W = \{(\mathbf{s'}_r^W)_{\xi_r} \quad (\mathbf{s'}_r^W)_{\eta_r} \quad 0\}^T \tag{6.45}$$

The magnitudes of the eccentricity vectors are evaluated as

$$e_r^V = \sqrt{(\mathbf{e}_r^V)^T \, \mathbf{e}_r^V} \tag{6.46}$$

$$e_r^W = \sqrt{(\mathbf{e}_r^W)^T \, \mathbf{e}_r^W} \tag{6.47}$$

The unit vectors, $\mathbf{n}_r^V$ and $\mathbf{n}_r^W$, are normal to the planes of contact at the points where the top and bottom journal bases touch the bearing wall. Referring to Fig. 6.14 these normal vectors are evaluated as

$$\mathbf{n}_r^V = \frac{\mathbf{e}_r^V}{\|\mathbf{e}_r^V\|} \tag{6.48}$$

$$\mathbf{n}_r^W = \frac{\mathbf{e}_r^W}{\|\mathbf{e}_r^W\|} \tag{6.49}$$

Referring to Fig. 6.14, the penetrations due to the contact between the journal bases and the bearing wall are calculated as

Fig. 6.14 Definition of the local coordinate system associated with the revolute clearance joint and the respective unit vectors

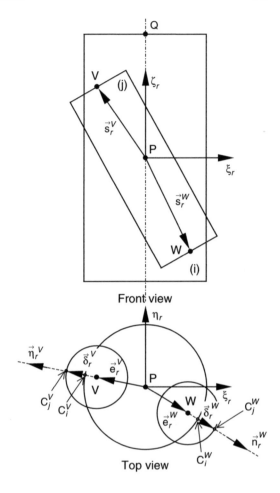

$$\delta_r^V = e_r^V - c \qquad (6.50)$$

$$\delta_r^W = e_r^W - c \qquad (6.51)$$

where e_r^V and e_r^W are, respectively, the modules of the eccentricity vectors at the top and bottom journal bases, and c is the radial clearance given by the difference between the radius of the bearing and the journal. There is contact if the radial motion exceeds the radial clearance size.

Considering C_i^V, C_j^V, C_i^W and C_j^W to be the potential contact points on bodies i and j, their global positions are evaluated as

$$\mathbf{r}_k^{C_k^V} = \mathbf{r}_k^P + \mathbf{A}_k \mathbf{s}'_k^{C_k^V}, \qquad (k = i, j) \qquad (6.52)$$

$$\mathbf{r}_k^{C_k^W} = \mathbf{r}_k^P + \mathbf{A}_k \mathbf{s}'_k^{C_k^W}, \qquad (k = i, j) \qquad (6.53)$$

where vectors $\mathbf{s}'^{C_k^V}_k$ and $\mathbf{s}'^{C_k^W}_k$, $(k = i, j)$, are the local coordinates of vectors $\mathbf{s}^{C_k^V}_k$ and $\mathbf{s}^{C_k^W}_k$ on bodies i and j, expressed by the local coordinate system of each body. These vectors are only defined in the local coordinate system of the joint $(\xi\eta\zeta)_r$ and are expressed as

$$\mathbf{s}'^{C_i^V}_r = \{0\ 0\ (\mathbf{s}'^V_r)_{\zeta_r}\}^T + R_i \mathbf{n}^V_r \tag{6.54}$$

$$\mathbf{s}'^{C_j^V}_r = \mathbf{s}'^V_r + R_j \mathbf{n}^V_r \tag{6.55}$$

$$\mathbf{s}'^{C_i^W}_r = \{0\ 0\ (\mathbf{s}'^W_r)_{\zeta_r}\}^T + R_i \mathbf{n}^W_r \tag{6.56}$$

$$\mathbf{s}'^{C_j^W}_r = \mathbf{s}'^W_r + R_j \mathbf{n}^W_r \tag{6.57}$$

The vectors defined by (6.54)–(6.57) when expressed in the local coordinate systems associated with bodies i and j yield

$$\mathbf{s}'^{C_i^V}_i = \mathbf{A}_{ri} \mathbf{s}'^{C_i^V}_r \tag{6.58}$$

$$\mathbf{s}'^{C_j^V}_j = \mathbf{A}^T_j \mathbf{A}_r \mathbf{s}'^{C_j^V}_r \tag{6.59}$$

$$\mathbf{s}'^{C_i^W}_i = \mathbf{A}_{ri} \mathbf{s}'^{C_i^W}_r \tag{6.60}$$

$$\mathbf{s}'^{C_j^W}_j = \mathbf{A}^T_j \mathbf{A}_r \mathbf{s}'^{C_j^W}_r \tag{6.61}$$

The impact velocities, required to evaluate the contact forces, using the Lankarani and Nikravesh model, are obtained by differentiating (6.52) and (6.53) with respect to time, yielding

$$\dot{\mathbf{r}}^{C_k^V}_k = \dot{\mathbf{r}}^P_k + \mathbf{A}_k \tilde{\boldsymbol{\omega}}'_k \mathbf{s}'^{C_k^V}_k, \quad (k = i, j) \tag{6.62}$$

$$\dot{\mathbf{r}}^{C_k^W}_k = \dot{\mathbf{r}}^P_k + \mathbf{A}_k \tilde{\boldsymbol{\omega}}'_k \mathbf{s}'^{C_k^W}_k, \quad (k = i, j) \tag{6.63}$$

The relative impact velocities between the two bodies at the contact points are

$$\Delta \dot{\mathbf{r}}^V = \dot{\mathbf{r}}^{C_j^V}_j - \dot{\mathbf{r}}^{C_i^V}_i \tag{6.64}$$

$$\Delta \dot{\mathbf{r}}^W = \dot{\mathbf{r}}^{C_j^W}_j - \dot{\mathbf{r}}^{C_i^W}_i \tag{6.65}$$

The relative velocities given by (6.64) and (6.65) are projected onto the direction normal to the penetration, yielding the relative normal velocities, $\dot{\delta}^V_r$ and $\dot{\delta}^W_r$, shown in Fig. 6.15. The normal relative velocities represent whether the contact bodies are approaching or separating. These velocities are evaluated by

$$\dot{\delta}^V_r = (\Delta \dot{\mathbf{r}}^V)^T \mathbf{n}^V_r \tag{6.66}$$

$$\dot{\delta}^W_r = (\Delta \dot{\mathbf{r}}^W)^T \mathbf{n}^W_r \tag{6.67}$$

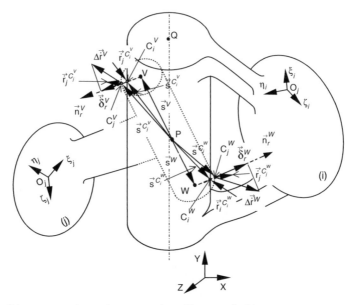

Fig. 6.15 Location of the contact points and representation of impact velocities

When contact between the journal and the bearing takes place, impact forces act at the contact points. The contributions of these impact forces to the generalized vector of forces are found by projecting them onto the X, Y and Z directions. Since these forces do not act through the center of mass of the bodies i and j, the moment components for each body need to be evaluated. For convenience and simplicity the bodies are presented separately in Fig. 6.16 and only the force components that act at the top journal base are illustrated. According to Fig. 6.16, the forces and moments working on the center of mass of body i are given by

$$\mathbf{f}_i = \mathbf{f}_N + \mathbf{f}_T \tag{6.68}$$

$$\mathbf{m}_i = y_i^{C_i^V} \mathbf{f}_i^z - z_i^{C_i^V} \mathbf{f}_i^y + x_i^{C_i^V} \mathbf{f}_i^z - z_i^{C_i^V} \mathbf{f}_i^x + x_i^{C_i^V} \mathbf{f}_i^y - y_i^{C_i^V} \mathbf{f}_i^x \tag{6.69}$$

The forces and moments corresponding to body j are written as

$$\mathbf{f}_j = -\mathbf{f}_i \tag{6.70}$$

$$\mathbf{m}_j = -y_j^{C_j^V} \mathbf{f}_j^z + z_j^{C_j^V} \mathbf{f}_j^y - x_j^{C_j^V} \mathbf{f}_j^z + z_j^{C_j^V} \mathbf{f}_j^x - x_j^{C_j^V} \mathbf{f}_j^y + y_j^{C_j^V} \mathbf{f}_j^x \tag{6.71}$$

Since the formulation of the spatial revolute joint involves a good deal of mathematical manipulation, it is convenient to summarize the main steps in an appropriate algorithm. This algorithm, presented in the flowchart of Fig. 6.17, is developed in the framework of the multibody methodology and can be condensed in the following steps:

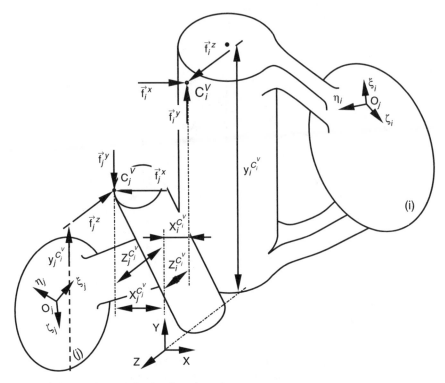

Fig. 6.16 Contact forces defined at the points of contact

1. Start at instant of time t^0, with given initial conditions for positions $\mathbf{q}^0$ and velocities $\dot{\mathbf{q}}^0$.
2. Define the location of points P_i, P_j, Q_i, V_j and W_j, necessary to describe the spatial revolute clearance joint. Define joint and material properties (R_B, E_B, ν_B, R_J, E_J and ν_J).
3. Compute the local coordinate system associated with the joint, $(\xi\eta\zeta)r$ and evaluate the unit coordinate vectors along each axis, that is, $\mathbf{u}'_\xi$, $\mathbf{u}'_\eta$ and $\mathbf{u}'_\zeta$ using (6.32)–(6.34).
4. Evaluate the local coordinates of the geometrical centers of the top and bottom bases in the $(\xi\eta\zeta)r$ system using (6.42)–(6.43).
5. Compute eccentricity vectors $\mathbf{e}^V_r$ and $\mathbf{e}^W_r$ and the unit vectors that define the impact direction $\mathbf{n}^V_r$ and $\mathbf{n}^W_r$ through (6.44)–(6.49).
6. Evaluate the penetrations δ^V_r and δ^W_r with (6.50)–(6.51).
7. Check for contact: if there is contact, determine the contact points using (6.52)–(6.53), evaluate the impact velocities with (6.66)–(6.67), compute the impact forces by (6.68)–(6.71) and add the impact forces to the equations of motion.
8. Obtain the new positions and velocities of the system for time step $t + \Delta t$ by integration of the final derivatives of the state variables.

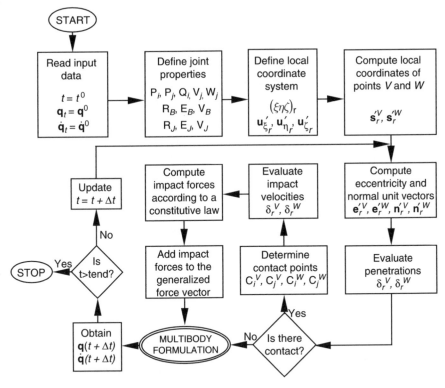

Fig. 6.17 Representation of the algorithm proposed to model spatial revolute joint with clearance in multibody systems

9. Update the system time variable.
10. Go to step 5 and repeat the whole process for the new time step, until the final time for the analysis is reached.

6.4 Demonstrative Example 1: Four-Bar Mechanism

In this section, the application of the four-bar mechanism that describes a spatial motion (Haug 1989) is employed, as an illustrative example to demonstrate how a spherical clearance joint can affect the behavior of the mechanism. The spatial four-bar mechanism consists of four rigid bodies that represent the ground, crank, coupler and rocker. The body numbers and their corresponding local coordinate systems are shown in Fig. 6.18. The kinematic joints of this multibody system include two ideal revolute joints, connecting the ground to the crank and the ground to the rocker, and one perfect spherical joint that connects the crank and coupler. A spherical joint, with a given clearance, interconnects the coupler and the rocker. This four-bar mechanism is modeled with 24 coordinates, which result from 4 rigid bodies and 19 kinematic constraints. Consequently this system has five degrees of freedom.

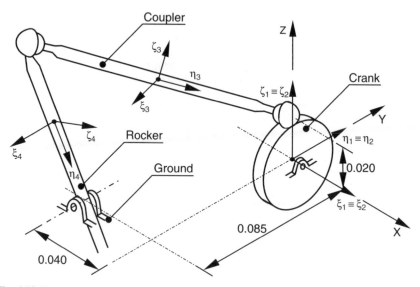

Fig. 6.18 Representation of the algorithm proposed to model spatial revolute joint with clearance in multibody systems

The initial configuration of the spatial four-bar mechanism is illustrated in Fig. 6.18. The system is released from the initial position with null velocities and under the action of gravity force, which is taken to act in the negative Z direction. The dimensions and inertia properties of each body are presented in Table 6.1. The dynamic parameters used for the simulation and for the numerical methods required to solve the system dynamics are listed in Table 6.2.

In order to study the influence of the spherical clearance model in the global behavior of the spatial four-bar mechanism, some kinematic and dynamic characteristics, corresponding to the first 2s of the simulation, are presented and discussed in what follows. The results are always plotted against those obtained with a simulation in which all kinematic joints are considered to be ideal or perfect.

The normal contact force and the joint reaction force, for the first impact at the spherical clearance joint, are shown in Fig. 6.19a. The plotted reaction force is the magnitude of the joint force in the revolute joint that connects the ground to the rocker. The simulation is performed by employing the Lankarani and Nikravesh contact force model given by (3.9). By observing Fig. 6.19a, it is clear how the impacts inherent to the dynamics of the clearance joint influence the reaction force.

Table 6.1 Geometric and inertia properties of the spatial four-bar mechanism

Body nr	Length (m)	Mass (kg)	Moment of inertia (kg m^2)		
			$I_{\xi\xi}$	$I_{\eta\eta}$	$I_{\zeta\zeta}$
2	0.020	0.0196	0.0000392	0.0000197	0.0000197
3	0.122	0.1416	0.0017743	0.0000351	0.0017743
4	0.074	0.0316	0.0001456	0.0000029	0.0001456

Table 6.2 Parameters used in the dynamic simulation of the four-bar mechanism

Socket radius	10.0 mm	Young's modulus	207 GPa
Ball radius	9.5 mm	Poisson's ratio	0.3
Radial clearance	0.5 mm	Integration step size	0.00001 s
Restitution coefficient	0.9	Integration tolerance	0.000001 s

The two force curves plotted show a very similar shape. The maximum reaction force is about 60% of the contact force. Figure 6.19b shows the hysteresis curves for the first three impacts at the spherical clearance joint. The contact force decreases for each impact suggesting that some system energy is dissipated from impact to impact. This dissipated energy is measured as the area enclosed by the hysteresis plot. The energy dissipation is due to the contact model used and, since the gravitational force is the only external action, that is, no other external forces or drivers are applied to the system, no energy is fed to the system.

Figure 6.20a–d depicts the Z-component for the position, velocity and acceleration of the center of mass of rocker, as well as the Y-component of the reaction moment that acts at the revolute joint that connects the ground to the rocker, for both ideal and spherical clearance joint simulations. The results plotted in Fig. 6.20a show that the position accuracy of the four-bar mechanism is clearly influenced by the effective joint clearance. Furthermore the maximum Z-position is not reached in every cycle since the impacts within the joint with clearance dissipate some of the system's energy. Figure 6.20c and d shows that the mechanism with clearance joint creates significantly larger dynamic accelerations and reaction moments on the system than those observed for an ideal dynamic model. The magnitude of acceleration and moment for the case of the ideal joint is very low, not even visible in the figures, since there is no driver in the system, the gravitational force being the only external action on the system.

The magnitude of the eccentricity vector is plotted in Fig. 6.21, in which the different types of motion between the ball center and the socket center can be observed,

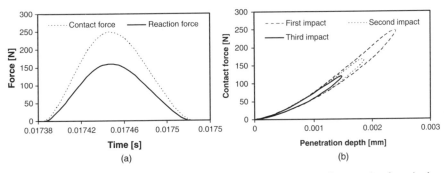

Fig. 6.19 (a) Normal contact force at the clearance joint and corresponding reaction force in the ground–rocker revolute joint for the first impact; (b) hysteresis loop of the first three impacts at the clearance joint. The contact force decreases from impact to impact because no energy is fed to the system

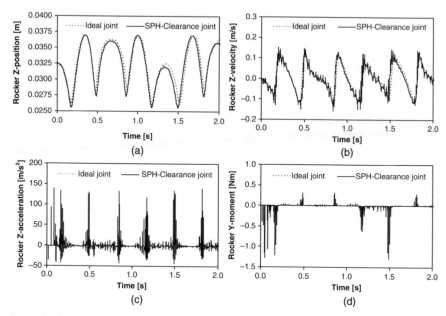

Fig. 6.20 (a) Z-coordinate of rocker center of mass; (b) Z-velocity of rocker center of mass; (c) Z-acceleration of rocker center of mass; (d) Y-component of the reaction moment at the ground–rocker revolute joint.

namely, free flight, impact, rebound and permanent or continuous contact. In the first instants of the simulation, free flight motion followed by impacts and rebounds are well evident. After that, it can be observed that the ball and socket present periods of permanent or continuous contact, where the ball follows the socket wall. The dashed line in Fig. 6.21 represents the radial clearance size (0.5 mm), which corresponds to the maximum relative motion between the ball and the socket without contact.

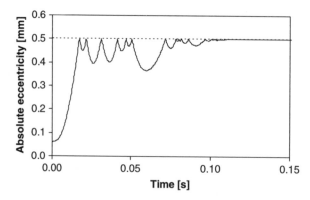

Fig. 6.21 Module of the eccentricity vector

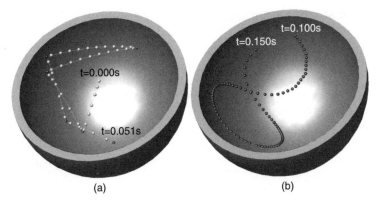

Fig. 6.22 Ball center trajectory inside the socket: (**a**) first simulation's instants in which free flight motion and impacts followed by rebounds are visible; (**b**) permanent or continuous contact, i.e., the ball follows the socket wall

The path of the ball center relative to the socket center is also illustrated in Fig. 6.22. Figure 6.22a shows the relative motion between the two bodies for the first six impacts. The half gray spherical surface represents the clearance limit while the small spheres inside represent the ball center path. The free flights are illustrated by clear spheres, whereas the impacts are represented by darker spheres. It is clear that in the first instants of simulation the impacts are immediately followed by rebounds. Figure 6.22b shows the time interval simulation from 0.100 to 0.150 s. From this figure, it is observed that the ball is always in permanent contact with the socket wall. Furthermore the permanent contact between the ball and socket is accomplished by varying the penetration depth along the radial direction. The Poincaré maps are used to illustrate the dynamic behavior of the spatial four-bar mechanism with a spherical clearance joint. The system's response is nonlinear, as the relative motion between the ball and the socket can change from free flights, impact and continuous contact, as illustrated in Figs. 6.21 and 6.22. The nonlinear system response is well visible by plotting the corresponding Poincaré maps, which relate the rocker Z-velocity versus rocker Z-position, shown in Fig. 6.23. The Poincaré map presented in Fig. 6.23b has a complex aspect, densely filled by orbits or points, which indicates chaotic behavior.

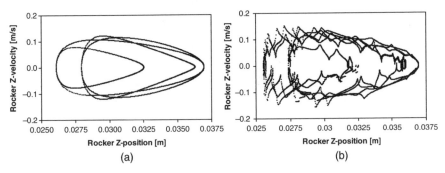

Fig. 6.23 Poincaré maps: (**a**) ideal joint; (**b**) spherical clearance joint

6.5 Demonstrative Example 2: Double Pendulum

In order to examine the effectiveness of the formulation developed for the spatial revolute clearance joint, a double pendulum with the configuration shown in Fig. 6.24 is studied. The numbering of the bodies of the system and their local coordinate frames are also pictured in Fig. 6.24. The double pendulum is made up of three rigid bodies, the ground body and two arms. One ideal revolute joint connects the two pendulum arms while a spatial revolute clearance joint, with a radial clearance size of 0.5 mm, exists between the ground and body 2. This simple multibody system is modeled with 18 coordinates and 11 kinematic constraints, which results in a system with 7 DOF.

Initially the double pendulum rests in the XZ plane position with pendulum arms perpendicular to each other. The system is then released from this initial configuration under gravity action only, which is taken as acting in the positive Y direction. The geometric dimensions and inertia properties of the double pendulum are listed in Table 6.3, while the dynamic parameters used in simulations are shown in Table 6.4.

In order to evaluate the influence of the spatial revolute clearance joint, in the dynamic performance of the double pendulum, the main kinematic and dynamic characteristics of the system response during the first 4 s of simulation are analyzed here, with the results compared to those obtained for a system with ideal joints. The

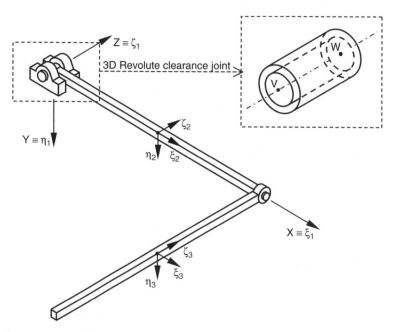

Fig. 6.24 Double pendulum modeled by two rigid bars and two revolute joints. Each bar is a prismatic homogeneous rigid body with square section of 0.03 m × 0.03 m

Table 6.3 Governing properties for the double pendulum

Body nr	Length (m)	Mass (kg)	Moment of inertia ($Kg\,m^2$)		
			$I_{\xi\xi}$	$I_{\eta\eta}$	$I_{\zeta\zeta}$
2	1.0	7.02	0.0010530	0.5855265	0.5855265
3	1.0	7.02	0.5855265	0.5855265	0.0010530

Hertz contact law with hysteresis damping factor, given by (3.9), is used to evaluate the contact forces caused by the impact in the clearance joint.

Figure 6.25a shows the normal contact force developed in the revolute clearance joint, during the first impact, and the reaction force of an ideal joint. The plotted reaction force is the module of the joint force in X direction developed at the ideal revolute joint that connects the two pendulum arms. In Fig. 6.25a, it is observed that the reaction force shape is similar to the shape of the contact force at the clearance joint. The maximum reaction force is about 50% of the contact force. Figure 6.25b shows the hysteresis curves for the first three impacts, developed at the clearance joint. As in the case of four-bar mechanism, analyzed in the previous section, the contact force decreases for each impact, suggesting that some of the system energy is dissipated from impact to impact. This dissipated energy is represented by the area enclosed by the hysteresis plot.

The position, velocity and acceleration of the center of mass of body 3 in the Y direction are plotted in Fig. 6.26a–c. Since the double pendulum has an open-loop topology, the existence of a clearance joint clearly influences the global position of the bodies of the system. The global behavior of the double pendulum with a clearance joint is characterized as nonlinear, almost chaotic, as illustrated in the Poincaré map of Fig. 6.26d. The velocity and acceleration components of the center of mass of body 3 in the Y direction are chosen to construct the Poincaré map. The complex densely filled appearance of the Poincaré map is an indicator that the system response is highly nonlinear.

The effect of the existence of a revolute clearance joint in the global motion of the double pendulum is illustrated in Fig. 6.27, in which the trajectory of the center of mass of the end arm is plotted during the first three seconds of simulation. The effect of the impacts, which occur at the clearance joint, produce very high peaks in the component of the reaction forces and moments used to represent the system response of the double pendulum, as it is shown in Fig. 6.28. Note that the reference quantities presented in Fig. 6.28, denoted as ideal joint, are obtained for the system model with ideal joints only.

Figure 6.29 shows the module of the eccentricity vector for both journal bases of the revolute clearance joint, as referred in Fig. 6.24. It can be observed that, for

Table 6.4 Simulation parameters for the double pendulum

Bearing radius	10.0 mm	Young's modulus	207 GPa
Journal radius	9.5 mm	Poisson's ratio	0.3
Radial clearance	0.5 mm	Integration step size	0.00001 s
Restitution coefficient	0.9	Integration tolerance	0.000001 s

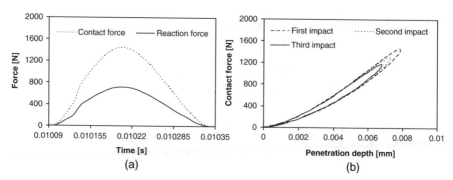

Fig. 6.25 (**a**) Normal contact force developed at the revolute clearance joint and reaction force in the ideal revolute joint that connects the two pendulum arms; (**b**) hysteresis loop of the first three impacts at the revolute clearance joint

the first impacts, the trajectories of the two bases are coincident. But, after that, the impacts between the journal and the bearing wall take place at different instants of time, meaning that some misalignment occurs. This phenomenon can be observed in Fig. 6.30 where a sequence of frames from a computer animation of the journal trajectory, relative to the bearing boundaries, is shown. The contact situations are represented by a dark journal, while the noncontact cases are represented by a light-colored journal. In the frame sequence the rebounds, during the first impacts, are

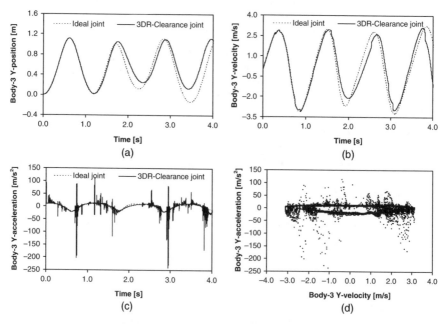

Fig. 6.26 (**a**) Y-position of body 3 center of mass; (**b**) Y-velocity of body 3 center of mass; (**c**) Y-acceleration of body 3 center of mass; (**d**) Poincaré map

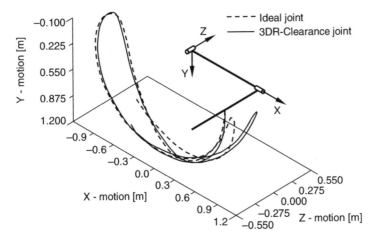

Fig. 6.27 Trajectory of the center of mass of the end arm during the first 3 s of simulation

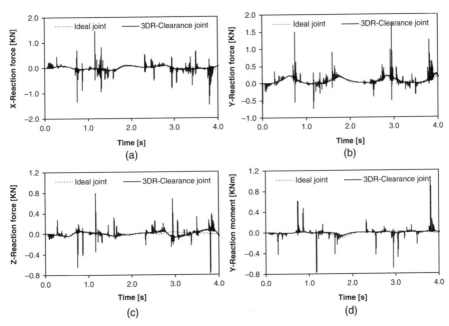

Fig. 6.28 Reaction forces and moment generated in the ideal joint: (**a**) X-reaction force; (**b**) Y-reaction force; (**c**) Z-reaction force; (**d**) Y-reaction moment

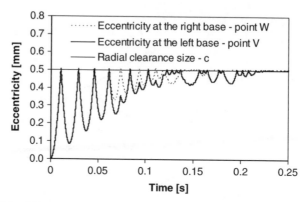

Fig. 6.29 Module of the eccentricity vector for the two journal bases of the revolute clearance joint

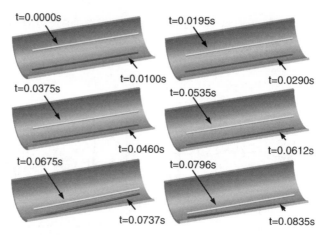

Fig. 6.30 Sequence of positions representing trajectory of the journal inside the bearing for first instants of simulation

clearly visible. Moreover the height of the rebound decreases from impact to impact due to the energy loss.

6.6 Demonstrative Example 3: Slider–Crank Mechanism

In this section, a spatial slider–crank mechanism is used as a numerical example to demonstrate the application of the methodologies previously presented. Four rigid bodies describe the slider–crank model under consideration. The model also includes one ideal revolute joint that connects the ground and the crank, one ideal spherical joint between the crank and the connecting rod and one ideal translational joint that connects the ground and the slider. A joint with clearance connects the slider to the connecting rod. This system has 19 independent kinematic constraints

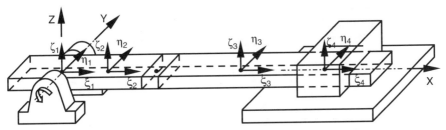

Fig. 6.31 Initial configuration of the spatial slider–crank mechanism with a clearance joint between the slider and the connecting rod

and it is described by 24 coordinates, which results in a mechanism with five degrees of freedom. A body-fixed coordinate system ($\xi\eta\zeta$) is attached to the center of mass of each body, and XYZ represents the global coordinate system. The slider–crank model shown in Fig. 6.31 is constrained to move in the XZ plane and therefore the overall motion described by the mechanism can be considered as planar. The gravitational acceleration is considered as acting in the negative Z direction.

The initial configuration is taken with the crank and the connecting rod collinear. The crank, which is the driving link, rotates about the Y-axis with a constant angular velocity of 150 rad/s. The crank velocity is maintained constant due to its very large rotational inertia, that is, the crank acts like a flywheel. The geometric characteristics and the mass and inertia properties are presented in Table 6.5.

The dynamic parameters used for the simulation are similar to those listed in Table 6.4. Furthermore the initial conditions necessary to start the dynamic analysis are obtained from kinematic simulation of a slider–crank model in which all the joints are considered to be ideal. In order to keep the analysis simple, all the joints are considered as frictionless.

In what follows, three different situations are analyzed. In the first one, the slider–crank mechanism is considered as a two-dimensional system and the joint between the connecting rod and slider is modeled as a 2D revolute clearance joint, as was presented in Chap. 4. In the second case, the slider–crank is modeled as a spatial multibody system and the clearance joint as a spherical joint with clearance, as developed in Sect. 6.2. Finally, in the third situation, the slider–crank is also considered as a spatial system and the joint clearance as a 3D revolute joint with clearance, as was presented in Sect. 6.3. For all three situations, the contacts between the elements

Table 6.5 Geometric and inertia properties for the dynamic simulation of the spatial slider–crank mechanism

Body nr	Length (m)	Mass (kg)	Moment of inertia (kg m^2)		
			$I_{\xi\xi}$	$I_{\eta\eta}$	$I_{\zeta\zeta}$
2	0.1524	0.15	10.0000	10.000	10.000
3	0.3048	0.30	0.0002	0.0002	0.0002
4	–	0.15	0.0001	0.0001	0.0001

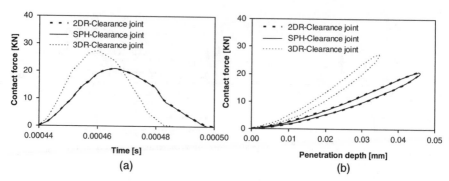

Fig. 6.32 (**a**) Normal contact force at the clearance joint for the different models; (**b**) hysteresis loop at the clearance joint. All the plots are for the first impact

that constitute the clearance joints are modeled by employing the Hertz contact law with hysteresis damping factor, expressed by mathematical equation (3.9).

The normal contact force developed at the clearance joint at the first impact, for the three different simulations, is plotted in Fig. 6.32a. Figure 6.32b shows the corresponding hysteresis curves. The contact force curves, obtained with 2D revolute clearance joint and with spherical clearance joint, coincide. This is expected

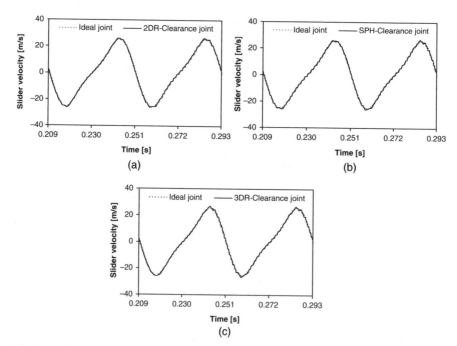

Fig. 6.33 Slider velocity for the different joint clearance models: (**a**) 2D revolute joint; (**b**) spherical joint; (**c**) 3D revolute joint

since the slider–crank model describes a planar motion and, therefore, the spherical clearance joint behavior can be considered equivalent to the case of the 2D revolute clearance joint. However, for the case of 3D revolute clearance joint simulation, the contact force curve presents a different evolution. Moreover the contact duration is shorter and the maximum force is greater than for the other two cases. This behavior can be understood due to the fact that in the 3D revolute model there are two point contacts, which are the two journal bases, instead of one point contact. Due to the nonlinear characteristics of the continuous force model, it is expected that representing the joint with one or two contact points leads to different results.

Figures 6.33 through 6.37 show the results for the case in which the radial clearance size is equal to 0.5 mm. In order to better understand the dynamic behavior of the slider–crank mechanism, these results are compared with those obtained for the ideal joint. The simulations are performed for the three different clearance joint models mentioned previously and at time interval corresponding to two complete crank rotations.

In Figs. 6.33 through 6.35, it is observed that the peaks on the slider acceleration curves are due to the contact force variation, which occurs during the period of contact between the elements that compose the clearance joints. The same phenomenon can be observed in the curves of the crank moment, because the contact forces are propagated through the rigid bodies of the slider–crank mechanism. It is noteworthy

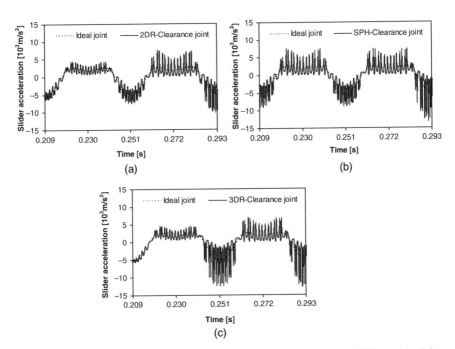

Fig. 6.34 Slider acceleration for the different joint clearance models: (**a**) 2D revolute joint; (**b**) spherical joint; (**c**) 3D revolute joint

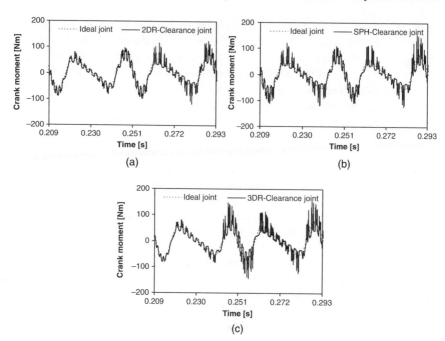

Fig. 6.35 Crank moment for the different joint clearance models: (**a**) 2D revolute joint; (**b**) spherical joint; (**c**) 3D revolute joint

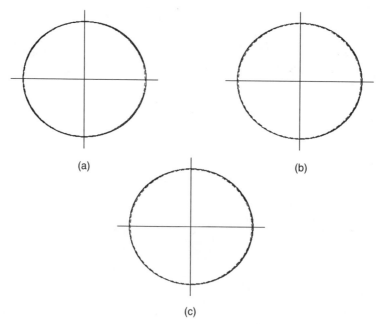

Fig. 6.36 Journal/ball motion inside the bearing/socket boundaries for different joint clearance models: (**a**) 2D revolute joint; (**b**) spherical joint; (**c**) 3D revolute joint

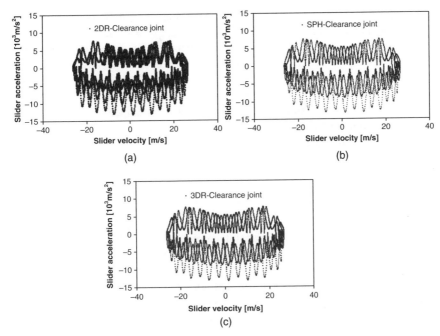

Fig. 6.37 Poincaré maps for the different joint clearance models: (**a**) 2D revolute joint; (**b**) spherical joint; (**c**) 3D revolute joint

that the global behavior of the slider–crank is the same for the three different clearance joint models, after the initial phase of the simulation. In fact, after $t=0.26$ s the magnitude of the moments is the same for all models. In the period prior to $t=0.26$ s, though the evolution of the crank moment is qualitatively the same, its magnitude differs for the different models. Figure 6.36 shows the relative motion between the journal and the bearing, and between the ball and the socket, for the three different joint clearance models. The dashed line represents the radial clearance size (0.5 mm). Since the slider–crank mechanism describes a planar trajectory, the relative motion between the journal and the bearing and between the ball and the socket occurs only in the plane XZ, the motion in the Y-direction being null. Therefore it is possible to compare the dynamic behavior of the 2D and 3D joint clearance formulations.

The dynamic response of the slider–crank mechanism is also represented by the evolution of velocity and acceleration of the slider and of the crank that acts on the crankshaft. Additionally the relative motion between the journal and the bearing, and between the ball and the socket centers, is plotted together with the corresponding Poincaré maps. The values of slider velocity and slider acceleration are plotted in the Poincaré maps in Fig. 6.37.

The Poincaré maps presented in Fig. 6.37 show that the global behavior of the slider–crank motion is nonlinear but tends to have a certain level of periodicity. Furthermore it is clear that both 2D and 3D models have a rather predictable motion, their correlation, when measured by the outputs plotted in the Poincaré maps, being very high. The dynamic response of the spatial slider–crank mechanism with both spatial clearance joints, presented in this chapter, is consistent with the results obtained for the planar slider–crank model analyzed in Chap. 3, in which the two-dimensional revolute clearance joint modeling was used. This is expected, since the formulation for the spherical clearance joints is quite similar to that of the planar revolute clearance joints. Moreover the spatial slider–crank mechanism used in this section, apart from being a three-dimensional model, describes a planar motion due to the constrained topology.

6.7 Summary

In this chapter, a formulation for spatial dynamic analysis of multibody mechanical systems, employing the Cartesian coordinates and the Newton–Euler's approach, was presented. Euler parameters were used to define the angular orientation of bodies, which leads to a mathematical formulation without singularities. Additionally a simple and brief description of the standard mechanical joints of spatial multibody systems was presented. The constraint equations for the perfect spherical joint and for the perfect three-dimensional revolute joint were also presented in the first section of this chapter.

A general methodology for dynamic characterization of mechanical systems with spherical and three-dimensional revolute joints with clearance was formulated, for implementation in general-purpose computer codes. This formulation can be understood as an extension of that proposed for the two-dimensional multibody mechanical systems with clearance joints. The descriptions of spherical and three-dimensional clearance joints are based on the Cartesian coordinates, the joint elements are modeled as contact-impacting bodies and the dynamics of the joints is controlled by a continuous contact force model, which takes into account the geometric and mechanical characteristics of the contacting bodies. The normal force is evaluated as a function of the elastic pseudo-penetration depth between the impacting bodies, coupled with a nonlinear viscous-elastic factor representing the energy dissipation during the impact process. For this continuous contact force model, it is assumed that the compliance and damping coefficients are available.

Three illustrative examples and numerical results were presented, the efficiency of the developed methodologies being discussed in the process of their presentation. In order to keep the analysis simple, the friction and the lubrication effects were not included in the present chapter. However, the inclusion of these phenomena closely follows the procedures described in Chaps. 4 and 5.

A spatial four-bar mechanism was used with a spherical clearance joint formulation to demonstrate its application. The system was driven only by gravity, and the system was not conservative due to the presence of damping in the impact model, which leads to some energy dissipation in every cycle of the motion. This was observed by comparing the position and velocity of the mechanical system with clearance to that of a system with ideal joints. Clearly the impacts within the clearance joint significantly increase the amount of dissipated energy.

In a second application, the double pendulum was used as a numerical example to illustrate the spatial revolute clearance joint formulation. In addition, a simple spatial slider–crank mechanism was used to study the influence of the clearance joint models in comparable planar and spatial mechanisms. A flywheel was incorporated in the crankshaft to maintain the crank angular velocity constant. It was observed that the overall results are consistent to those obtained for planar slider–crank model, that is, due to the clearance impacts the dynamic system presents much higher peaks in the acceleration time response and reaction forces than would be predicted if clearances were neglected. The system behavior clearly tends to be nonlinear and eventually chaotic, as shown by the corresponding Poincaré maps.

The overall results presented in this section show that the introduction of clearance joints in spatial multibody mechanical systems significantly influences the prediction of the components' position and drastically increases the peaks in acceleration and reaction moments at the joints. Moreover the system response clearly tends to be nonlinear when a clearance joint is included. This is a fundamental feature mainly in high-speed and precision mechanisms where the accurate predictions are essential for the design of the mechanical systems.

References

Baumgarte J (1972) Stabilization of constraints and integrals of motion in dynamical systems. Computer Methods in Applied Mechanics and Engineering 1:1–16.

Haug EJ (1989) Computer-aided kinematics and dynamics of mechanical systems. Vol. I: basic methods. Allyn and Bacon, Boston, MA.

Lankarani HM, Nikravesh PE (1990) A contact force model with hysteresis damping for impact analysis of multibody systems. Journal of Mechanical Design 112:369–376.

Nikravesh PE (1988) Computer-aided analysis of mechanical systems. Prentice Hall, Englewood Cliffs, NJ.

Nikravesh PE, Chung IS (1982) Application of Euler parameters to the analysis of three-dimensional constrained mechanical systems. Journal of Mechanical Design 104:785–791.